OUT OF NATURE

Out of Nature

Why Drugs from Plants Matter to the Future of Humanity

KARA ROGERS

THE UNIVERSITY OF
ARIZONA PRESS
TUCSON

THE UNIVERSITY OF
ARIZONA PRESS

www.uapress.arizona.edu

Library of Congress Cataloging-in-Publication Data

Rogers, Kara.
    Out of nature : why drugs from plants matter to the future of humanity /
Kara Rogers.
        p.   cm.
    Includes bibliographical references and index.
    ISBN 978-0-8165-2969-8 (pbk. : alk. paper)   1. Pharmacognosy.   2. Natural
products.   3. Drug development.   I. Title.   II. Title: Why drugs from plants
matter to the future of humanity.
    RS160.R64   2012
    615.1′9—dc23   2011032771

17   16   15   14   13   12   6   5   4   3   2   1

All drawings are by the author unless otherwise noted.

Contents

Illustrations

OUT OF NATURE

Prelude

PEEKING OUT FROM THE starboard side of the jib, I could see another 5-foot wave rolling toward us. I shifted my weight out, leaning back over the water to level the heel we took on as a nearly 30-knot gust filled the sails and began to push the boat over. I braced myself for the cold spray that would come shooting up from the bow. Glancing toward the stern, I saw my husband, Jeremy, extend his body way out over the water too. After a moment, the gust broke and the boat leveled. Then the wave came, and the spray hit. The water chilled me to the bone.

The sun had long ago disappeared behind a never-ending stream of gray clouds. Cool gusts of air had turned sailing conditions from swift and sure to unruly and foaming. We flew down the backside of the wave. I shifted my weight to level the hull as the wind steadied again. It was blowing somewhere around 20 or 22 knots now. As we dipped down into the trough, I prepared myself for the next wave. The water looked rough and as though it stretched endlessly beneath the turbulent, clouding sky. It rolled and chopped along. In some places, it looked green, in others black, effects of the light and cloud shadows cast down from above and the dilution of color from the upswell of sand below. The water never looked the same. From moment to moment and day to day its appearance was always changing. We rode up and over another wave.

"Ready to tack," my husband called out. "Ready," I answered. A hesitant moment passed. "Tacking."

I pulled the jib line out of its cleat; the boom began to swing. I crawled as nimbly as I could to the other side of the centerboard, pulled on the opposite jib line, and cleated it. Half a second later we were positioned on the smooth curled edge atop the boat's hull, our feet secured under the hiking straps, our bodies straining to balance the 14-foot-long vessel as the wind pushed us through the water. I looked back at our wake. It was large for such a small boat. And for a brief moment, I let myself revel in the undying grace of the water and sky surging on the horizon behind us. Then I turned my attention to what lay to the south, off our port side, settling my eyes again on the horizon. The skyline of Chicago dominated in the distance, and Lake Michigan looked like a small ocean. As long as we stayed near land and tacked back and forth on a beam reach, we would be alternating between feeling isolated and distant on rough waters in one direction and existing amid millions of people in the third most densely populated US city in the other direction.

My favorite moments are sailing away from civilization. Living among so many bodies, the desire to seek solitude can be very strong. The city is fantastic because of convenience, and I like to think of it as a center of controlled chaos. During the week, humanity rushes forth to its heart. On the weekends, people linger in the parks and take their time moving from point A to point B. Beyond the city lie the suburbs, miles and miles of homes, neatly mowed lawns, and an SUV or a minivan in every driveway. America's suburbs are famous for landmarks driven by compulsive consumerism—Target; WalMart; Home Depot; Bed, Bath, and Beyond; and the like. After living in a city like Chicago, most modern-day suburbs are comparatively devoid of character. The outlying residential communities of the city, just like those of Los Angeles, Houston, Phoenix, Baltimore, and most other American cities, are the epitome of the relentless pursuit of the "American dream." And beyond many of America's suburbs, the land characteristically opens into broad patches of crops and pastures filled with various livestock in what we like to call rural America. Farmhouses sit surrounded by farm machinery on hundreds of acres of farmland, sometimes rolling, sometimes as flat as a plate.

Beyond the rural lands, there exist the remnants of wild lands. This is the American wilderness, and included among its many relatively untouched features are canyons, mountains, prairies, and forests.

Much of our country's wilderness is protected in national parks and reserves, now major tourist destinations. Countries from Indonesia and Southeast Asia to Central and South America have equivalent wildernesses, encompassing a diverse range of ecosystems and land-forms, including rain forests, savannas, deserts, and long chains of mountains. Many of these areas are rich in biodiversity, character-ized by the vast number and variety of species they contain. Many also may contain large numbers of yet-unidentified plants, insects, and other creatures.

There have been many times when we are sailing away from the city and toward open water that I find myself wondering what the lake will be like in ten or twenty years. Will the choppy water in front of us still seem as fickle and wild as it does now? Or will it shimmer with plastic bags and bottles? Lake Michigan is no stranger to the destruction caused by human activity. Water runoff from farms and cities, pollu-tion from watercraft with two-stroke engines, and the introduction of nonnative species, including the zebra mussel, which has devas-tated native clam populations and wreaked havoc on power plants and water intake systems, have plagued the lake for decades. These factors, which have beset all the Great Lakes of North America within the last century and a half, have caused the lakes' ecosystems to change.

Ecologists are only just beginning to understand how to quantify and accurately predict the long-term consequences of such dramatic and swift distortions. Their models for study are ecosystems of the world already very nearly destroyed as a result of human activities, from deforestation to the introduction of nonnative species. The ability to describe ecosystems in economic terms is one of the great-est challenges facing conservation. The economic impact caused by the introduction of the zebra mussel into the Great Lakes serves as an important example. Rebuilding water intake systems to prevent clogging from mussels, clearing mussel shells from shores, and deal-ing with other mussel-related problems have already cost millions of dollars and are expected to cost the surrounding regions billions more in the coming years.

Harder to predict, however, are the economic and social impacts arising specifically from the loss of species. The destruction of habi-tat, which inevitably leads to declines in the diversity of plants and animals, can produce a rippling effect across an ecosystem, causing

dramatic and potentially irreversible shifts in predator, prey, and pollinator populations. The loss of habitat and the extinction of species
also affect us. We value nature for its aesthetic qualities. We venture
into it in search of recreation and relaxation. The loss of the animals
and plants that populate the natural areas we visit decreases the aesthetic value of these areas. But far more complicated is comprehend-
ing the impact of species loss on the functioning of human societies
and on human health and medicine. Our daily lives and health are
intertwined with the well-being of our environments.

Factors in this equation that are not often discussed are the consequences on medicine rendered by the loss of species, by declines in
Earth's biodiversity. The extinction of plants and animals represents
the extinction of potential medicines. Most new drugs are modeled
on compounds discovered in nature, and many of these natural compounds come from plants. But for the last several decades, the discovery of new drug compounds has been left largely to technology—to
methods of chemical synthesis that have produced far fewer successful drugs than expected.

Natural-products drug discovery is an area of research in which,
like so many other pursuits related to drug development, there are
profits to be made. Preservation of nature is a comparatively less
expensive endeavor, and it too is profitable. But because its greatest
profits are determined over the long course of time, rather than from
one fiscal quarter to the next, and because its benefits are far more
abstract than hard market figures, it is often perceived as economically limited. Biomass, soil generation, aesthetic value, and other services that ecosystems provide us have not historically carried a price
tag. Their true value, however, is much, much more than what we
currently place on them as societies.

Nature is a world of stark contrasts and harmonic forms to which
we seem instinctively drawn. We go to great lengths to see nature
and to re-create our experiences with it. Plants, which form our
gardens and enrich urban landscapes, play an enormously important
role in supporting the natural habitats we enjoy visiting so much. In
the wild, in their natural homes, plants give life to the soil, provide
homes and shelters for animals, and endow the world with beauty,
inspiring us artistically and spiritually. But plants, like many animals,
are declining in their diversity. To see them disappearing from the

wild, and largely as a result of our actions, is extraordinarily difficult to accept.

The loss of species is a symptom of the modern *Homo sapiens'* disconnect with nature. The current rate of human-caused environmental destruction is staggering. It is made all the more astonishing when one considers the limited availability of most natural resources. But there are many contradictions between the way nature is viewed on personal, industrial, and governmental levels. Researchers may support conservation efforts, and companies may publicly declare their support for green technologies. The government may enact laws to protect the environment. But the amount of money invested on private and federal levels is often a gesture aimed at acknowledging public sentiments concerning the environment, rather than an amount sufficient to actively work to protect it.

When it comes to nature conservation, natural-products drug discovery can play a vital role. It also has a role in protecting the knowledge and culture of indigenous peoples, who often want to and should be involved in natural-products discovery. But negotiating prior informed consent and benefits sharing between companies and indigenous tribes is a delicate process, and many companies have shied away from natural products as a result. Now, however, finding solutions to issues concerning the "ownership" and conservation of nature has become necessary to the advancement of drug development. Without solutions, new medicines to cure cancers and to treat some of humanity's worst afflictions will remain ideas and goals explained on paper.

The many advances in technology, from electricity and automobiles to planes, computers, and cell phones, have made survival for the majority of humans alive today far easier than it was for humans alive at any other time in history. But technology alone will not prevent the loss of animals and plants. We must reconnect with nature, with the world that ultimately defines our existence and produces our foods and medicines. Indeed, the tiny pills that we hold in our hands and that heal our ailments originated, in most instances, not from humankind, but from the natural world. They are fundamentally *out of nature.*

In writing and illustrating this book, I found myself perpetually amazed by plants and their influence on our lives. It is my hope that others, in reading this book, will feel similarly inspired.

I

Plants and Medicine

Robbins' cinquefoil (*Potentilla robbinsiana*).

IN 1992 THE US FOOD AND DRUG ADMINISTRATION (FDA) approved an agent called Taxol for the treatment of ovarian cancer. In the years that followed, the list of cancers found to be susceptible to Taxol grew significantly, so that by the early 2000s, it was also being used to fight malignant disease of the lungs, breast, and head and neck. It was saving lives, and it became one of the most important anticancer drugs ever discovered.

But it was not always such smooth sailing for Taxol. In fact, several times in the course of its development—thirty years from discovery to market—Taxol very nearly sank. From the damp, muted rain forests of the Pacific Northwest to the bedsides of patients suffering from malignant disease, the winds that filled Taxol's sails and propelled it to success were continually unpredictable, gusting one moment and calm the next. The whole of Taxol's relationship with nature in the past half century was marked by capricious patterns of excitement and promise and despair and anger. The events of its past reflect a very real dilemma that travels quietly in the background of drug discovery and conservation—the saving of human lives at nature's expense.

The story of Taxol began in 1962, when botanist Arthur Barclay, working for the US Department of Agriculture (USDA), found himself staring up at a hillside of Pacific yews in Gifford Pinchot National Forest, one of the oldest protected forests in the United States. One of Barclay's final tasks in a four-month-long field expedition to the West was to collect samples of the Pacific yew tree, *Taxus brevifolia*. Barclay found himself in the forest near the base of Mount Saint Helens on a warm August day and in the company of three graduate students, who helped him gather stems, fruit, and bark. A few days later, Barclay shipped the dried specimens to the USDA office back east. Extracts from the samples were screened in search of new chemical compounds that had the potential to be developed into drugs. In early tests, bark from the Pacific yew proved promising.

Barclay returned to Pinchot forest in 1964, this time collecting some 30 pounds of yew bark. The specimens were again sent back to the USDA headquarters. This time, however, they were forwarded to the National Cancer Institute (NCI), to Monroe Wall, a scientist who was overseeing work on a compound known as paclitaxel (or Taxol, as Wall referred to it), which had been isolated from yew samples during the initial screening. Wall was at the then new Research Triangle Institute in North Carolina. Prior to his position there, he had worked for the USDA, where he assisted in the screening of plant compounds. He was known for his skill as a medicinal chemist and was given the Taxol project knowing that he was one of few who possessed the knowledge necessary to characterize the compound's

A young Pacific yew (*Taxus brevifolia*). (Photo credit: Jeremy D. Rogers)

structure and activity. But while Wall and his colleague Mansukh Wani had succeeded in rapidly identifying the antitumor properties of Taxol, they struggled to elucidate the compound's structure. In 1971 they finally published a report describing the structure and antitumor characteristics of Taxol. They then decided to hand the project over to other NCI scientists for testing in animals. But the research came to a grinding halt.

The compound's structural complexities were massive, which complicated its study. Perhaps more significantly, however, was that Wall was able to extract only half a gram of Taxol from every 30 pounds of bark. Because of the difficulty of the extraction process and the severe limitations in its natural source, the NCI sidelined Taxol. Still, curiosity about the compound persisted, and several scientists continued to investigate its anticancer properties. In 1979 a report describing its mechanism of action against cancer cells served to reinvigorate interest in Taxol. The groundbreaking work was reported by Susan Horwitz, a researcher at Albert Einstein College of Medicine in New York City.

Prior to the work of Wall, Wani, and Horwitz, the Pacific yew and its close relatives in the *Taxus* genus weren't known to be of any value medicinally. The Tsimshian, an indigenous people of the Pacific Northwest, however, knew otherwise. The Pacific yew shares a long history with the Tsimshian and other Native American peoples of the Northwest coast, who often carved the trees into canoes and furniture. The Tsimshian in particular, however, were well aware of the medicinal properties of the tree, long before the NCI researchers ever were. They rendered medicines from the bark that were used to treat a variety of ailments. But most surprising is that the Tsimshian appear to have produced a remedy from the tree's bark specifically for the treatment of cancerous diseases. Of course, whether this treatment was truly effective is unclear.

The Pacific yew is a slow-growing species. A century may pass before it reaches the stage where its bark can be harvested for Taxol. A century is longer than most people's lives, and it's about as long as the modern pharmaceutical industry has been in existence. Compounding this was the problem that one tree supplied one dose of Taxol, which meant that the value of the drug was limited by the availability of its natural source. Despite this, in the late 1980s, the bark was stripped off yews as though there were an endless reservoir of them, and suddenly Taxol's reputation as a wonder drug began losing ground to concerns about yew conservation. Harvesting the bark of yews meant cutting the trees down, and at the rate the trees were disappearing, it wouldn't take very long to completely extirpate the species.

The threat to the Pacific yew evoked strong reactions, especially since the trees were being scraped from the land in a part of the country known for its old-growth forests and native wildlife and for its environmentalist activities. The Pacific yew and other trees found in old-growth forests in the Northwest provide vital habitat for the endangered northern spotted owl. In 1990 *Strix occidentalis caurina* came under the protection of the Endangered Species Act, with the result that the owl's native habitat—the national forests of Washington and Oregon—was rendered off-limits to the timber industry.

The battle over the use and protection of forests is, of course, nothing new to the Northwest. The debate between preservation and protection to keep land available for logging and other uses began long ago. The different sides were represented by natural-ist John Muir and US forestry leader Gifford Pinchot. Muir advo-cated for federal forest preservation, particularly in northern Pacific regions of the United States. Pinchot argued for sustainable forest use, often hedging in favor of the timber industry. But in 1990, when the northern spotted owl and its habitat fell under federal protection, a move that would have pleased Muir, members of the timber indus-try were enraged.

The "cancer versus owl" debate created tension throughout com-munities in the Northwest. What was more important: saving human lives or saving a species from extinction? Even the most avid envi-ronmentalists, if they had loved ones dying from cancer, would have found it difficult to oppose yew harvesting. Fortunately, there was one way to effectively circumvent the problem. The owl's sur-vival clearly depended on yews and old-growth habitat. Taxol's survival, on the other hand, did not necessarily depend on the har-vesting of entire yew trees. Synthetic chemistry offered a way out. Given the mounting tensions, if scientists were to continue their pursuit of the drug, they would need to learn more about the com-pound's structure. They would need to find a way to generate it synthetically.

In the early 1980s it was virtually impossible to even in part re-create synthetically the chemical reactions that produce Taxol. In France, Andrew Greene and Pierre Potier were investigating the leaves of multiple plants in the *Taxus* genus. Their investigations revealed chemical precursors that in the laboratory could be sub-jected to processes that ultimately produced Taxol. But the process was extraordinarily inefficient. In the late 1980s, US scientist Robert Holton developed a synthetic process to create taxusin, a compound closely related to Taxol. This discovery led to his subsequent devel-opment of a semi-synthetic technique for Taxol that was much more efficient than the process developed by Greene and Potier. The tim-ing of Holton's work was crucial. The NCI was under tremendous pressure to produce greater and greater quantities of Taxol for inves-tigations of the drug. This in turn had increased demands on yew harvesting. Following the approval of Taxol in 1992 for the treatment of ovarian cancer, Holton further improved his synthetic technique,

The native habitat of the Pacific yew (*Taxus brevifolia*), shown in shaded regions. (Source data: U.S. Department of Agriculture)

enabling cost-efficient extraction of starting material from the renewable resource of yew needles. The advance served to significantly reduce tensions over yew-harvesting initiatives and allowed for the production of greater quantities of Taxol than could be extracted from yew bark. Although total synthesis of Taxol likely will never be possible, due to the cost of materials and the complexity of synthesis, partial synthesis using needles rescued the Pacific yew from overharvesting and eliminated at least one threat jeopardizing the habitat of the northern spotted owl.

The story of Taxol and the Pacific yew is one that has been discussed in settings as diverse as graduate-level pharmacology courses and public meetings on environmental issues. The story is so significant because both environmentalists, seeking to protect the Pacific yew and northern spotted owl, and pharmaceutical researchers, seeking to produce Taxol more efficiently, were triumphant. It was a rare victory for medicinal plants, and it represented a crucial turning point in the relationship between drug discovery and the environment.

Drugs like Taxol serve as bridges between humanity and nature. In our modern world, it is very easy to forget how closely tied to the natural world we really are. So much of our existence is spent interacting with technology that we increasingly find ourselves indoors rather than out. Our priorities are continually shifting toward technology, and in some cases, these shifts are arguably not for the better. Enslaved by television and the Internet, it is easy to lose sight of the importance of unplugging ourselves, of going outside and exploring nature. Our youngest generations are already experiencing some unwanted side effects of such indoor-oriented societies; not only are a third of all children in developed countries suffering from obesity, but they also have been raised in a sheltered world, where home heating, air conditioning, and the availability of drugs that can heal us are too readily taken for granted, and where playing in the dirt and literally coming into contact with nature are viewed as opportunities to acquire some rare disease, not as opportunities to learn.

As I became increasingly curious about the human relationship with nature and our passion for technology, I began to wonder about pharmaceuticals. Just about every one of us relies, at some point in our lives, on a drug to prevent disease or to restore or maintain our health. But we have an oddly paradoxical perception of modern drugs—we are simultaneously distrustful of the pharmaceutical industry and dependent on the agents it produces.

Drugs form the center of conventional medicine, but they also are important components of alternative and traditional medicine. These three systems, long isolated from each other by their unique histories and precepts, have now found themselves very much closer, united by the precarious future of nature. Loss of natural habitat and decreases in populations of wild medicinal plants have narrowed the selection of species available to treat disease. Likewise, the current state of drug development is rather tenuous, particularly because there is a lack of diversity in the agents making their way through the pharmaceutical pipeline. In drug development, just as in living biological systems, diversity is fundamental to survival.

The ecosystem of the pharmaceutical industry is one characterized by companies that develop unique, patentable compounds. Profits on patented agents provide money for investment in research and development of more compounds. So, profits go hand in hand with patents, and these days it seems that every substance with any remote pharmaceutical potential is patented.

But the metrics used to determine the "potential" of compounds seem remarkably distorted. The work of natural-products discovery lies in academia, where researchers work to answer fundamental questions about newly discovered compounds. Basic research is time consuming and expensive, and pharmaceutical companies cannot always afford to carry out basic research. Rather, they often partner with academia, absorbing novel compounds and putting the most promising ones on the fast track to clinical trials. In some instances, to minimize costs, a drug's development is expedited, generally to the exclusion of thorough exploration of its mechanism of action. In fact, knowledge about the mechanism by which a drug exerts its therapeutic actions in the human body isn't necessarily required for approval; the drug need only produce a beneficial effect and have minimal side effects. Knowledge of how a drug works at cellular and molecular levels, however, can provide valuable information about cell dysfunction in disease and explain an agent's beneficial and harmful effects, so a great deal of effort goes into elucidating these properties even after an agent is approved.

The processes driving drug development on industrial scales have been repeatedly refined, so that now each company can tailor its work specifically to addressing what is asked of it by entities such as the FDA. The FDA traditionally has held high standards in its drug-approval process, although it experienced a brief lapse in the mid-1990s and early 2000s. After a period of intense criticism, the lingering effects of which are still rippling through the agency, it appears to be again holding pharmaceutical companies to rigorous approval standards. Yet, drugs continue to barely clear the investigative hurdles that must be passed to bring them to market. Development is a cost-versus-benefits process. If a compound derived from an already-existing drug gives any slight indication that its profits will outweigh the financial input needed to support its research, the pursuit for exclusivity and development accelerates, guided all the while by efforts to maintain opportunities for maximum profit margin in the end. A casualty of this process is the potential for complete novelty, the search for substances entirely new to medicine, substances that could be the answers to our most pressing disease problems.

In recent years, patents held on drugs manufactured by pharmaceutical companies have been expiring left and right. Between 2011 and 2016, some $255 billion in sales are expected to disappear as a result of blockbuster drugs losing patent protection, leaving companies scrambling to find new drugs to replace the ones that once brought in huge profits. While there are new, modified compounds shuffling in to replace the old ones, novelty—the discovery of compounds with new core structures—is very rare. Instead, the general approach has been to find new applications for already marketed drugs.

Of all biomedical endeavors, drug discovery and development are among the most costly and the most time consuming, and to make matters more difficult, costs are increasing annually. In 1976 the average total expense for drug research and development (R&D), which includes discovery, biochemical analysis, testing in animals, and clinical testing in humans, was about $54 million. By 1996, these costs had skyrocketed to more than $500 million, and by 2001 they had jumped again to $800 million. These days, it takes at least seven years and costs more than $1 billion to bring a single new drug from discovery to market. The primary driving force for increasing expenditure was the introduction in the 1990s of more stringent regulations, forcing pharmaceutical companies to spend more money on clinical trials, which provide the most useful information in terms of establishing profiles of drug safety and efficacy.

In recent years, however, the pharmaceutical industry has witnessed a dwindling yield in commercial agents per dollar invested. Most drug candidates fall out of the running during bioactivity screening. Those that survive screening are put to the test in clinical trials, which further narrow the field. A significant portion of the remaining few are cut out of the picture at the federal approval level, usually due to side effects deemed too risky or too frequent to be accepted. The industry is also facing other, perhaps more significant problems, ones that are causing the pipeline to dry up altogether. Many of these issues revolve around the approach to drug discovery and development, hinging largely on concerns in pharmaceutical philosophy and in biotechnology.

There are multiple approaches to drug development. Most of the compounds pursued by pharmaceutical companies have broad marketability and can benefit large numbers of people. Pharmaceutical companies thrive on this approach because it yields blockbuster drugs—any agent that brings in $1 billion or more in revenue in one year for the company that owns it. Blockbusters are a rare breed, however. To find them, many companies have come to rely on a carefully calculated approach, the design of drugs that target specific molecules in cells, particularly individual genes and proteins that go awry to give rise to very specific types of conditions, such as a certain form of lung cancer or chronic nerve pain. For some of these drugs, genetic testing can indicate whether and how well a patient will respond to them, and this in turn has initiated a move toward personalized medicine, in which doctors can determine the most effective therapeutic strategies for their patients on the basis of one's genetic makeup.

For optimal treatment of patients, personalized medicine is the pinnacle of success. For drug companies, tailored drugs are not as profitable and might even be more expensive to develop than agents generated using traditional approaches. The increased expense and risk of development of these drugs are due largely to the fact that each personalized drug must be accompanied by its own diagnostic test to identify individuals in whom the drug can and cannot be used. To meet these demands, many drug companies have had to partner with biotech companies or develop their own diagnostic R&D branches. Progress has been slow. The hope, however, is that personalized medicines, prescribed on the basis of genetic compatibility, might not only improve disease treatment but also work their way into disease prevention regimes, which would expand the market to people beyond those already affected by disease. In the long run, preventing diseases, so they never have the opportunity to cause a sudden trip to the doctor, can save trillions of dollars in medical expenses.

The development of personalized drugs, however, is a field riddled with controversy, not only because of the expenses associated with their development, which translates into increased expenses for health-care providers and consumers, but also because these agents will work only in selected groups of people, leaving everyone else wondering when their personalized drugs will become available. Also, as more companies shift toward the development of population-specific drugs, interest in channeling efforts into the development of drugs that are relevant on a broader scale could decrease. Antibiotics, antivirals, vaccines, and other broadly applicable agents represent groups of drugs for which the world is desperate. Their development, however, is not very enticing for private companies. These types of agents must have low manufacturing costs so countries worldwide can afford them, and they must work reliably, stemming the spread of disease before it escalates to epidemic or pandemic levels. The combination of these factors alone presents a daunting problem. Faced by a paucity of novel antimicrobials, our ability to treat the world's most common infectious diseases disappears altogether.

Diversity among drugs, with regard to not only the small or large populations they target but also the range in their individual chemical structures, is important for many reasons. For instance, the development of agents that can be used to treat individuals who experience life-threatening side effects when they take otherwise commonly prescribed drugs has saved many lives. One example is anaphylactic shock resulting from an allergic response to penicillin. If it were not for the development of non-penicillin-derived antibiotics, there would be no form of treatment for bacterial infection in the subset of people who experience this reaction. However, if the agents that hit the market in the first decade of the 2000s were any indication, drug diversity has atrophied. This lack of diversity has been fueled by the "monoculturing" of drugs—essentially the introduction of minor chemical modifications to slightly alter the activity of old drugs, thereby propagating families of agents. This strategy has served drug development well, but in recent years it has become one of the primary mechanisms deployed for the generation of "new" agents.

"New" in the sense that, even though small modifications are made in the chemical structures, all the agents still possess the same core structures, and therefore most are limited to nearly the same applications, consequently threatening our ability to defend against disease. An obvious analogy is relying on a single type of plant as a source of food. Genetically similar plants, grown across a large expanse, are rendered uniformly susceptible to devastation by disease, drought, or other factors. While some years may produce bumper crops, it only takes one year of devastation to create widespread crop failure.

In the mid-twentieth century, scientists working to identify many novel, diverse compounds from natural entities were the first to experience the ups and downs of modern drug discovery. The compounds discovered in the 1950s and 1960s were fed into laboratory studies and, in some instances, into trials in humans. Much of this early research was fueled not only by scientific curiosity but also by the hope that compounds capable of treating the world's worst afflictions would be found. Substantial progress was made toward the latter goal, considering that some of the drugs that emerged during this period of pharmaceutical development—agents such as penicillin and the antituberculosis drug isoniazid—have greatly reduced human suffering. Interestingly, however, the concept of blockbuster drugs did not exist then, and while money was important, it was not the only force driving discovery.

During that era, most drug research was performed at university labs or at government research facilities, where large screening projects were undertaken. Scientists explored a wide variety of substances, resulting in the identification of a number of compounds with real potential for drug development. For every compound with potential, however, there were many that failed to demonstrate significant activity against human disease or that defied investigation with the technologies available at the time. In the end, the freedom of scientific exploration was reined in. Maintaining massive screening projects to identify potential drug compounds proved too expensive. In the United States, the primary screening project, the National Cancer Institute's Developmental Therapeutics Program, which was initiated in the 1950s, fizzled out in the 1980s, having been whittled down and truncated by budget cuts that eventually undermined the effectiveness of the research process. The truth about drug discovery and development—that it is expensive and time consuming—took a long time to be realized. When the money ran out, the fun and excitement of the novelty of isolating and characterizing compounds practically vanished. A larger force was needed to move compounds from the laboratory to the pharmacy. That force proved to be private industry and, specifically, big pharma.

As government-run drug-discovery programs disappeared, private companies took over. With the ability to develop and market commercial drugs came fundamental changes in the aims of drug discovery. Today, drugs are commodities, boiled down to expenses and profits. Companies are engaged in constant battle against increasing costs, driven by a plethora of factors capable of instantly running up research and manufacturing expenses. Within this plethora, standing out like an elephant among mice, is technology. There are a number of technologies at the disposal of the pharmaceutical industry, and a great deal of time and effort is spent in implementing and optimizing them. Some of the more revolutionary tools that have emerged in the last half century include combinatorial chemistry, automated sample processing and screening technologies, and genomics. To say that such technologies represent a departure from the past is an understatement. More accurate is to say that they have severed drug discovery from its roots.

In the early twentieth century the discovery of new compounds to be tested for potential pharmaceutical development relied solely on the identification and isolation of chemicals from biological sources, such as plants, animals, and microorganisms. As science advanced, it became possible to use these chemicals as lead compounds for the subsequent generation of synthetic and semi-synthetic compounds. In many cases, synthetically derived compounds are designed to be more potent and effective, as well as safer and more practical for commercialization, than their naturally occurring parent chemicals.

A single natural product can spawn multiple daughter compounds, as illustrated by the discovery of penicillin and the later synthesis of methicillin and related antibiotics. The story of penicillin and its introduction to medicine also reflects a very real element of science— serendipitous discovery. In 1928 Scottish scientist Sir Alexander Fleming accidentally discovered the antibacterial effects of a substance secreted from the mold *Penicillium notatum.* He confessed to the accidental nature of his discovery in 1941, in a letter to the *British Medical Journal,* when he wrote about his finding: "It is true that all the work on this substance originated in the accidental contamination of a culture plate with a spore of this penicillium, and in my first paper on the subject a photograph appears of this culture plate."

Fleming had a knack for stumbling across monumental findings. In the opinion of some, he was a tinkerer, someone who thought it worth taking the time to explore and experiment with potentially unfruitful endeavors, to do things that other scientists would have thought ridiculous. In the opinion of others, he was just plain lucky. His groundbreaking contributions to science appear to have been a combination of both, although tinkering, intuition, and curiosity undoubtedly combined to set him up for discovery. In 1921 while working in the laboratory he inadvertently contaminated a culture of bacteria with his own nasal mucus as a result of having a runny nose from a cold. He knew that the plate was contaminated, but rather than throwing it out, he decided to keep it and observe it. This led to his discovery of lysozyme, an enzyme produced by the body that plays an important role in defending against bacterial infection. Knowledge of the existence of this enzyme opened new doors to the study of the human immune system. Today scientists continue to study lysozyme, particularly in the contexts of cancer, infection, and autoimmune disorders.

Fleming's discovery of lysozyme was huge, but it was his discovery of penicillin that changed the world. The ultimate medical value of penicillin was realized through the work of Oxford University scientists Howard Walter Florey and Ernst Boris Chain, who eventually isolated and purified penicillin from the mold in the 1930s, a feat that Fleming was himself unable to accomplish. The discovery by US scientist Mary Hunt in the early 1940s of another *Penicillium* species, *P. chrysogenum,* which was capable of producing twice as much penicillin as *P. notatum,* enabled the substance to be generated on a commercial scale. Hunt's discovery was reportedly prompted by the arrival of Florey and a colleague by the name of Norman Heatley in the United States prior to the onset of World War II in Europe. The men brought with them their cultures of *P. notatum.* US scientists attempting to up the production of penicillin by the mold were frustrated by the large fermentation tanks needed to grow enough of the organisms to generate substantial quantities of the drug. Hunt, who worked at the government-funded Northern Regional Research Laboratory in Peoria, Illinois, where the scale-up for penicillin production was taking place, went in search of moldy fruits and vegetables to see if she could identify another type of *Penicillium*. And she did. Her resourcefulness and determination were remarkable, and because of her efforts, mass production of the drug was initiated in time for the war.

If penicillin had been developed today, it would be a blockbuster. It was first given to soldiers during World War II and later came into widespread use among the general population. But regardless of revenue, the discovery of penicillin was significant because it marked the beginning of the antibiotic revolution, a period in the history of drug development when the power of pharmaceutical agents to treat deadly or previously incurable bacterial diseases was realized on a global scale. It was not long after penicillin came into public use, however, that resistant bacteria began to emerge. Structural diversity is the key to preventing the emergence of resistant organisms, but this was only beginning to be understood at the time.

Between the 1930s and 1960s, much was learned about the chemical structures of compounds, and new physical chemistry methods were devised, whereby scientists could more easily determine the atomic and molecular arrangements of chemicals. At the same time, there also occurred major advances in biochemistry, which improved scientists' understanding of how cells and molecules function and interact and enabled the more precise measurement of chemical effects on cells and the human body. These advances contributed to the production of second-generation antibiotics, including methicillin, which is a semi-synthetic agent that was developed in the late 1950s in response to the emergence of penicillin-resistant bacteria. Although resistance to methicillin was observed by the 1960s, resulting in its discontinued use, its generation ignited an increase in interest in the development of synthetic drugs.

Synthetic drug development in the twentieth century actually began with the work of German chemist Friedrich Wöhler, who in the 1820s synthesized urea, a breakdown product of proteins in animals. Wöhler's discovery led to rapid advances in chemical synthesis.

A century later, German researcher Paul Ehrlich, along with Japanese scientist Hata Sahachiro, synthesized a series of compounds from arsenic and tested them for activity against the bacterium that causes syphilis. This work led to their discovery of arsphenamine (Salvarsan), the first synthetic compound derived from a parent compound, in this case arsenic. Ehrlich's concept of drug development, in which he envisioned the modification of chemicals via synthetic processes and the subsequent testing of these compounds for biological activity, laid the foundation for modern drug discovery. Using this approach, chemists would be able to generate a large variety of compounds. Ehrlich's work ushered in dramatic shifts in the overall approach to drug development. For thousands of years, humans relied on natural entities as sources of medicines, and until the early twentieth century, the majority of medicines that humans used for the treatment of diseases and ailments came from plants.

The takeover of drug discovery by chemical synthesis is a fairly recent event, beginning in earnest in the 1920s and 1930s, when scientists studied the chemical structures of compounds isolated from plants and attempted to re-create the structures synthetically. Scientists added different side groups to various places on the structures of the natural compounds, thereby generating a number of derivatives, all of which had in common the same basic molecular organization of the naturally occurring structure. All therefore belonged to the same drug class. These new derivatives were test-driven in the laboratory to see if they exhibited antimicrobial effects and to determine whether they possessed other biological activities.

In the case of certain classes of antimicrobial agents, because there is such distinct similarity in the basic molecular arrangement between natural compounds and their synthetic derivatives, an organism that is resistant to one of these drugs is usually resistant to all its structural relatives. Consequently, new original chemical structures are needed to create a broader arsenal of antimicrobials. Plants and other organisms are valuable sources for antimicrobial substances because they naturally produce chemicals to kill off specific predators, including bacteria, fungi, and protozoans, some of the major groups of human microbial pathogens.

Perhaps, in the discovery of novel compounds, a return to nature is in order. With the rise of synthetic drug development, scientists came to rely more on their ingenuity in the laboratory and less on rooting around in nature to discover new compounds. And as the synthesis of compounds became increasingly sophisticated during the twentieth century, and as the number of synthetic compounds grew tremendously, there was born the pharmaceutical industry as we know it today. Libraries of compounds, many of which contain structures with slight variations on some previously discovered natural product, were generated, and the process of screening these compounds for biological activity accelerated. However, one facet of this process that has changed very little in the last century is the fact that out of the large number of compounds screened, only very few are suitable drug candidates. And when one candidate emerges and its biological effects are elucidated, there often ensues an effort to identify and generate related synthetic compounds based on that candidate's structure.

This frenzy of activity led to the generation of important drugs, including the anticancer agents letrozole and tamoxifen, the HIV/ AIDS drug zidovudine (AZT), and the group of cholesterol-lowering drugs known as statins. The pharmaceutical industry relied heavily on a number of technologies to generate these drugs, though at the heart of the basic structures of many of them are the structures of natural compounds. The significance of this is frequently diminished by the excitement that a new technological approach actually worked, perpetuating the belief that modern drug development can be an entirely lab-oriented system, that it can operate independently of the world that exists on the opposite side of the window.

Pharmaceutical companies are designed for rational, high-throughput drug development. This is the fastest way of identifying suitable candidate compounds and of generating drugs in a profitable manner. But how heavily should we rely on synthetic technologies to *advance* drug discovery? A tremendous amount of effort has gone into streamlining drug development, and the technologies implemented over the course of the last several decades are important. But they are not working as productively as they were expected to. The failure of combinatorial chemistry and other technologies to ramp up the production of new agents has led to a shortage of novel parent compounds and other new agents in the commercial pipeline. The consequences of this are now being felt across the pharmaceutical industry.

All the work that went into improving synthetic drug development beginning in the 1970s and 1980s meant that there was little time left for the discovery of novel compounds in nature. In fact, nature no longer seemed to be relevant. Older scientists detached from natural products and young researchers rarely realized the importance of staying connected to nature, especially as more and more new—and unproven—technologies were developed. The emergence of these new approaches meant that there was more preliminary work to be done in the lab, and it needed to be done before any actual study of drugs could be performed and the results taken seriously.

From the safety and comfort of the laboratory there emerged an unrealistic perception of human health, the notion that we can exist in isolation, apart from all other life on Earth. But in the twenty-first century, as species discovery has regained significance as a worthwhile endeavor, scientists have also realized that plant compounds and other naturally occurring substances still hold a great deal of promise for medicine. Scientists suspect that there are very many novel bioactive compounds from natural products remaining to be discovered, so many, in fact, that the task at hand is extremely daunting. Testing thousands and thousands of unknown compounds translates into billions and billions of dollars spent on substances that scientists can make little or no prediction about concerning whether or not they will ever see the light of day in marketed drugs.

Tied to the demands of turning profit into new tools and applications, scientists have been left with little or no time to plan drug-discovery expeditions. Furthermore, funding to support such efforts is sparse. For companies, sending researchers out to study rare plants and to collect specimens that might contain new compounds valuable to drug development has long been considered an unnecessary expenditure of both time and money, which in part explains why many companies now rely on others to discover new natural products. Despite the contagious pursuit of proving the latest technological advancements, there are some academic scientists who have secured funding to travel to places such as rain forests and other habitats rich in biodiversity where they are able to conduct research on plants and plant compounds. Government-funded bioinstitutes have also been established in multiple countries, enabling researchers from near and far to study and characterize local flora and fauna.

New plant compounds that are discovered can be sold to or shared with pharmaceutical companies, which then investigate the substances for their potential as lead compounds or commercial agents.

The International Cooperative Biodiversity Groups (ICBG), an effort supported by the US National Institutes of Health and National Science Foundation, has attempted to unite drug discovery with the preservation of natural resources and traditional knowledge. The latter element, the traditional knowledge of plant medicines, connects natural-products drug discovery to the history of humankind. The history of our species is punctuated by the things that drive human behavior and that influence the growth of civilizations and human thought. Factoring the diversity of humans and the histories and traditions of our species into the process of drug discovery can have a lasting influence.

Plants have served as sources of medicines and nutrients for humans for millennia. Some 11,000 years ago, our species began the transition from hunter-gatherer food acquisition to the domestication of animals and plants. The cultivation of plants, especially grains, was one of the first major steps toward freeing members of human societies from the constraints of basic survival. Planting, growing, and harvesting crops required attention and time for only several months out of every year. Although much time still needed to be devoted to raising families, building shelters, and other life tasks, there was suddenly more time available for new endeavors, many of which were oriented toward technology or artistic expression. Concerning the former, not all societies advanced at the same time or the same pace. By today's standards, some have not advanced significantly at all. One explanation for these differences, and one that presents the most likely scenario, was offered by scientist Jared Diamond in great detail in his book *Guns, Germs, and Steel*. Diamond suspects that domestication and the technological advancement of human societies were influenced primarily by geography and hence climate and environmental factors.

But plants are more than just sources of food to humans. They are gifts of nature. Their bright colors and diverse shapes and sizes make them interesting to observe and study, and their complex chemistries make them valuable sources of substances that heal our diseases. Among the most ancient plant-based systems of medicine that are still widely practiced today are Ayurveda, which originated in India, and the well-known traditional method of medicine in China. Ayurveda is believed to have come into existence about 3,000 years ago, and traditional Chinese medicine, though rooted in written tales and stories approximately 4,000 years old, has been practiced for some 2,500 years. Prior to the emergence of these systems of medicine, there was the *Ebers Papyrus*, a record of the herbal medicines used by the physicians of ancient Egypt. This manuscript, which dates to 1550 BCE, is considered to be one of the oldest medical works known. In fact, the system of medicine practiced by the ancient Egyptians is believed to date to at least 3000 BCE, making it a full 2,000 years older than Ayurveda.

In many of the earliest systems of ancient medicine, the relationship between plants and humans was largely spiritual in nature. The history of Ayurvedic medicine is embedded in stories of Hindu gods and sages, determined to end the suffering endured by the people of their lands. The wise gods, who were in possession of the knowledge and understanding of life and healing, passed on to others, presumably sages, who were genuine in their desire to help others, the body of knowledge that is Ayurvedic medicine. The sages themselves sometimes then became gods. The knowledge handed down appears to have been first recorded in the four Vedas, essentially collections of poems, hymns, chants, and incantations. Each Veda is an exposition of what had been previously transmitted to future generations in an oral tradition. The collection is filled with spiritual and natural accounts relating to Ayurvedic medicine, and among these stories of healing are descriptions of plants and various human ailments.

The importance of plants to ancient Ayurvedic practitioners is evident in various passages in the Vedas. In one scene, a physician says to a plant: "You are supreme." The Vedic texts also describe a number of human afflictions, including dropsy, leprosy, cough, fever, dysentery, malaria, skin diseases, and heart conditions. At the same time, the texts bring elements of the natural world into a supernatural context. Passages within these texts are quite interesting and reveal practical insight into human disease.

The *Atharva Veda* contains a hymn about a plant used to treat what is translated as leprosy. But only recently have scientists confirmed that leprosy was indeed a disease of ancient times. A 4,000-year-old skeleton discovered at an archaeological dig at Balathal, a site located in the present-day northwestern Indian state of Rajasthan, revealed evidence of infection with *Mycobacterium leprae*, the organism that causes the disease. Joint degeneration and lesions in various bones closely matched similar evidence from skeletons dating to the Middle Ages that were known to be affected by leprosy. Balathal itself dates to the Chalcolithic period, or Copper Age, which began about 3700 BCE. Prior to the discovery of the Balathal skeleton in 2009, there was doubt surrounding the reference to leprosy in the hymn in the *Atharva Veda*, as well as to a mention of the disease in the *Ebers Papyrus*. Now, however, it appears that the ancient Egyptian and Indian physicians did indeed recognize the disease, and they attempted to treat it.

Today, the acquisition of medicinal plants in places such as India, China, and Latin America has grown complicated. India, for example, is associated with three of the world's thirty-four recognized biodiversity hotspots, which are regions defined by a high degree of existing endemic, or native, biodiversity but, due to severe habitat loss, are at high risk of losing a substantial percentage of their wealth of life-forms. In the Himalayas in particular, overharvesting of medicinal plants has been cited as a significant contributing factor to the disappearance of native vegetation. In countries worldwide, overgrazing of livestock, deforestation, and crop farming have stripped away native vegetation. The Mesoamerican hotspot, an area extending from central Mexico to the Panama Canal, has been undergoing steady deforestation and other forms of habitat destruction since the 1800s, with the result that today only an estimated 20 percent of the original habitat remains intact. The area is home to about 24,000 different known types of plants. It is difficult not to wonder how many species have been lost over the last two hundred years, plants that scientists never even had a chance to discover, plant compounds forever unknown.

Our current relationship with nature is a bit of a calamity, and, as evidenced by places like the Himalayas and the Mesoamerican hotspot, our activities are certainly having an impact on natural-products drug discovery. The absence of promising new antibiotics and other drugs and the spread of epidemic and pandemic diseases are issues encrusted in environmental peril, the destruction of habi-tats, and the loss of irreplaceable animals and plants. The issues sur-rounding drug discovery and nature encompass a variety of scientific and human concerns that touch on technology, instinctual human behaviors, and the environment. The first two of these—technology and human behavior—have significantly influenced the vanishing connection between humanity and nature, which over the course of the last several hundred years has contributed to the rise of some of the most troubling environmental concerns that we face today.

Plants have existed on Earth far longer than humans have. They are masters of aquatic and terrestrial environments, and they live in harmony with animals, sometimes in ways that very nearly compro-mise their own survival. And for all the physical differences that we are able to discern between ourselves and plants, we have more in common with them than we might think. Plants are fundamental to our survival, a factor that defines our relationship with them, a relationship that has been misinterpreted of late, as reflected in our behavior toward them.

2

Humans and Plants

Bristlecone pine (*Pinus longaeva*).

ONE OF THE MOST INSPIRING things about plants is their indi-
viduality. Each species has its own distinct set of features, such as a
flower unique in scent or color or a fruit distinct in taste. Even the
same plant—a tree in the front yard or a shrub by the window—is
never quite the same from year to year. The often cyclical and predict-
able lives of plants—the breaking of buds, the blooming of flowers,
the falling of leaves—remind us that nature is not static. Constant
change defines nature, whether that change is growth, death, or adaptation. The emergence of a tiny garden plant from the turmoil of earth
brings hope. A human hand planted the seed, a human tended the
soil, and so a human is in one way or another connected to this plant.
Our relationship with plants is complex, both collectively as a species
and individually as people, and in our modern world, understanding
and defining this relationship have become increasingly challenging.

Trees, the most noble members of the plant kingdom, are symbols
of strength and wisdom. They are living history, built upon trunks
that tell the stories of their lives. For some trees, such as bristlecone
pines, these stories may be very long, with thousands of chapters,
one for each year of a tree's life. One of the oldest bristlecones, an
individual of the species *Pinus longaeva*, is nearly 5,000 years old.
This particular tree is so ancient that it has been named Methuselah.
Methuselah the tree, however, has lived almost five times longer
than his biblical counterpart of the same name, and his home is
located in an undisclosed area in the White Mountains of California.
The tree has experienced the greater part of an era in the Pacific
coast's history marked by human habitation, from the first native
tribes to occupy the land to the pressing expansion of modern cities.

There are trees in the Americas that have seen Christopher Colum-
bus come and go and that live to tell us, in the patterns of their rings,
what the weather was like and how old they were when the admiral
disembarked from the *Niña*. A tree's growth rings tell us when life
was good, when it was able to grow quickly and set down a new,
thick layer of xylem, the inner water-conducting tissue of plants that
in trees forms the sapwood and the heartwood. The rings tell us, too,
when life barely progressed, when basic survival, rather than growing taller and wider, was the primary concern. We can tell, from the
patterns, shapes, spacing, and nicks in the rings, if and when factors
such as fire, wind, drought, and even infestation with insect larvae

Bristlecone pines are some of Earth's longest-lived species. The
Rocky Mountain bristlecone pine, *Pinus aristata*, is shown. (Photo
credit: Jeremy D. Rogers)

influenced a tree's life. In this way, trees are portals to long-forgotten
eras in the history of both humans and the environment.

Deciduous trees remind us that Earth works in cycles. The shading
of leaves from green to yellow and red in autumn tells us that winter
is right around the corner. The cyclical patterns that characterize

the lives of plants have been repeated year after year ever since the first vascular plants appeared on Earth, more than 410 million years ago. And each year since modern society developed in the United States, complete with trains and automobiles, the onset of fall colors in northern regions has served as a famous attraction, drawing visitors from all over the country and the world. In the United States, the changing of the leaves is such a popular event that the Forest Service maintains a phone hotline for information on "fall foliage hotspots"— the national forests that are the most popular for viewing fall colors. Of course, the phenomenon of fall colors extends far beyond North America, to wherever deciduous trees grow, and in many of these places, leaf-peeping tourists are an annual economic boon.

At its most fundamental level, the changing of leaf color is a process of senescence and heralds the onset of winter dormancy. In woody plants, in the final days before leaf fall, an abscission layer forms at the base of the petiole, or the stalk of the leaf. This layer enables a clean break between the leaf and the branch, which prevents the underlying cells from being exposed to frost or potential disease-causing entities. The colors that leaves acquire in autumn are a function of plant pigments and biochemistry. As the number of hours of sunlight decreases steadily through autumn, and as temperatures cool, the production of chlorophyll, which gives leaves their typical green color, drops. This occurs in all deciduous species, including aspens, maples, oaks, birches, and hickories, among others. As chlorophyll fades away, other pigments in the leaves, namely the carotenoids and anthocyanins, become more prominent, giving rise to yellow, orange, brown, and red colors. Each deciduous species takes on a characteristic hue, and collectively, within a forest, these hues produce a vivid palette of color. Sometimes, where stands of different species congregate, these colors become heavily interwoven, making for a mixture of shades across a wide spectrum. In other instances, where individuals of a single species live clustered together, there occur loud splashes of color—a pool of yellow or red or orange.

The color that the leaves of a tree adopt is a product of differing leaf biochemistries. For example, the leaves of some trees contain high concentrations of glucose, whereas those of other trees contain high concentrations of waste products. The effects of these differences are illustrated by the golden yellow leaves of certain beech trees versus the brown leaves of some oaks. The golden shade of the beeches arises as a result of the combined presence of tannins and yellow carotenoids in the leaves, whereas the brown of the oaks is due to the presence of overwhelming quantities of tannins. Oak leaves are thought to contain such high levels of tannins because the chemical is toxic to animals that might eat the leaves.

Precisely when leaves change color and begin to fall depends on a tree's genetic programming. The tree's goal is to harness as much energy from the sun for as long as possible through autumn without sustaining damage from frost. Some trees can hold out for much longer than others, but the line between maximum nutrient acquisition and risk of damage from frost and disease is very narrow. The capturing of nutrients until the last possible moment is of vital importance for plants that must endure long winters. The more nutrients that are stowed away in a plant's inner tissues, the better its chances for survival and for getting a head start on growth in the spring.

But if fall colors really only mark the onset of leaf senescence, why do people go to such great lengths, sometimes flying overseas or driving for an entire day, to see them? There are probably many answers to this question. The simplest explanation, however, likely lies in the calm and warmth of the palette of vivid earth-tone colors that suddenly erupts and lasts at its peak for only a few days. Seeing the ephemeral event reminds us that nothing is forever, that even as nature is producing such wonderful sights to behold, it is in the process of taking them away, to leave behind the bare, protruding branches that announce the arrival of winter.

Plants dominated Earth long before humans ever appeared, and they did so without the ability to move under their own power. Imagine being anchored to one place for your entire life. From birth, every plant holds a bit of soil. This is its home, its own place in the world. Plants cannot see or hear, at least not in the way that we can see and hear. Rather, they experience life mainly through chemical communication. The ability of green plants to use the sun as a source of energy not only supports the plants' own survival but also that of all the life-forms around them, including us. Plants depend on carbon dioxide, which the animals around them provide in adequate quantities.

But there are many things about plants that we do not fully understand. For instance, plants do not "feel" animals eating their leaves, but chemical signals sent from the grazed parts tell them that something is there, and chemicals can be produced and released in toxic quantities as a mechanism of defense. Yet, herbivores are often permitted to nibble here and there. In many cases, only once an animal starts munching away too vigorously will a plant begin to produce a toxic chemical to warn them off. Is this because, as foliage is reduced, the chemicals become more concentrated within the plant? Or do plants possess some other mechanism, something within that respects the fact that they depend on the carbon dioxide and wastes of animals and must to some degree support animal survival?

As with all other life-forms on Earth, plants are compelled to reproduce, an activity that can be carried out by a variety of means, depending on the type of plant. In fact, of all biological systems on Earth, plants possess the greatest diversity of reproductive strategies. For instance, the reproductive systems of flowering plants may be described by any of a number of different terms, depending on whether male and female reproductive parts occur together on the same flower, separately on different parts of the same plant, or separately on different individual plants. The latter, dioecious plants, are very much like ourselves, with each plant being either male or female.

Flowering plants are those species characterized by the growth of flowers and often some sort of fruit that contains seeds, which are then dispersed on the air or are eaten by animals and carried off and deposited elsewhere. Eventually, in some cases within days and in other cases after many years of trying, a little seedling will sprout up from the soil. The seedling represents a plant's life goal. Translated from plants to humans, with the exception that we are mobile, the underlying threads of plant life change very little when compared with human life. We depend on plants for oxygen, we cooperate with nature as much as we need to, and we reproduce to ensure that our family lines are carried on in future generations.

While the applications we've devised for plants provide a general outline of how we interact with them, it is the basic similarities in nature that make our relationship with plants so fascinating. Plants have skin, circulatory vessels, cells, DNA, and male and female reproductive cells. They communicate with one another and with other organisms through chemical and mechanical reception, much the same way that animals communicate through smell, touch, taste, sight, and sound.

Some of the most intriguing comparisons of humans to other living forms were made by the ancient Greeks, who seemed to specialize in drawing heavily from nature for most of their philosophical arguments. Empedocles, who lived between 490 and 430 BCE, described as roots what we today know to be the components of matter. He insisted in his philosophy of physics that "roots" can be combined to produce all the different things we see in nature. Many of his theories make use of similar botanical metaphors. He saw likenesses where many people now fail to see the point of the exercise. A truly insightful notion was his use of plants to describe facets of human biology, right down to the development of the human embryo. Empedocles believed that humans go through a process of development that parallels that of plant seedlings, which bud and then grow, eventually gaining their full height and developing leaves and flowers. He concluded that humans, in their very first stages of development, must also grow from seeds. To make this analogy, Empedocles would have needed not only to understand how plants grow and develop but also to see similarities, however peculiar they may seem, between this process and that of human development.

Empedocles, similar to other natural philosophers of his time, also considered the soul as it relates to nature and the cosmos. He believed that the roots of life and the soul are virtually interchangeable, and in his poem "On Nature," the soul is examined in the cosmos as a whole. As such, the cycle of life is a part of the soul, and since plants have roots and life cycles, it followed that they have souls too. But the question of whether or not plants have souls is actually far more complex than Empedocles considered, depending on how one draws the line between having and not having a soul. It was Aristotle's thoughts on these issues that seem to have really mattered. He considered plants to be somewhere between having souls and being soulless. This placement of plants between the animate and inanimate seemed to appease conflicting arguments about the plant soul, at least up until the Middle Ages and the popularization of the doctrines of Christianity, which split souls into the material and spiritual. Then, in many belief systems, the human soul became known as the rational soul, beasts possessed animal souls, and plants possessed vegetative souls. Based on medieval conjectures stemming from the philosophical tenets of Scholasticism, of the three types of souls, it was only the rational soul that was considered capable of joining in union with God. Only the rational soul could be immortal. This system of beliefs came to dominate human life in Europe in the Middle Ages. At that point, for many people, the rational soul was the only soul that mattered—life itself became oriented around the task of making it into heaven, or, rather, avoiding hell.

The material souls given to plants and animals extinguished at physical death. So, material souls did not matter when it came to heaven and hell. Abiding strictly by the precepts of Christianity, and by the philosophies of Scholasticism and dualism, one may, with many assumptions and exclusions, be able to make sense of this reasoning. Yet, religious beliefs diverge substantially when it comes to discussing the nature of souls. And further problems arise within philosophies centered around dualism, which lacks a satisfactory explanation for mechanisms by which an immaterial form, namely the spiritual soul, by definition not a part of the physical world, is able to dictate the activities of a material body that is understood by modern science to be controlled by biological factors. Plants and animals, including humans, all are beings whose lives are the products of physical reactions, from conception to death. All life on Earth is made up of the same basic elements, and living organisms generally fulfill some biological purpose that supports other organisms, forming, quite literally, a circle of life. So, what is the line of reasoning that makes humans so different from other organisms? Perhaps more to the point, what makes humans *think* that they are more important than the rest of Earth's living beings?

In the modern era, the debate about whether plants are intelligent forms of life with souls reemerges periodically, if for no other reason than to reaffirm already existing dogma. But contrary to what many people may realize, there exist a diversity of beliefs about plants and souls. What is perhaps the most interesting facet of this is the complex reasoning that underlies these beliefs. In Jainism, for example, there exists living essence, or jiva, which is characterized by three components: happiness, consciousness, and vigor. Jiva is associated with a body and fills the space within the body. The living essence can be mobile or immobile, and thus bodies of the earth, the water, and even the fire and air are considered jiva. For plants, this means that they, and their parts—their roots and seeds—have souls. Under this orthodoxy, even bacteria and grains of sand have souls.

Few people, however, regardless of religious beliefs, are willing to attribute soullike characteristics to plants. The main problem is the issue of consciousness. Plants do not have brains and thus are not capable of thought in the same way that animals are. It is this distinction that raises humans above plants in the hierarchy of life. But just like humans and other animals, each plant is an individual, distinguished by what we see in them—an unusual knot in a tree's bark, an odd twist in a stem, a shrub's rebellious determination to take over a neighbor's fence. Plants live by nature's rules of survival. They stay alive as long as it takes to ensure the success of the next generation. This basic tenet is shared by plants and animals, and it is perhaps the most defining characteristic of life on Earth. And by this property alone, we have far more in common with plants than most people acknowledge.

Plant anatomy and physiology are really quite amazing. There are some 380,000 described species of plants in the world, and they all share the ability to capture sunlight and convert it into chemical energy through photosynthesis. The land plants that we see around us today also share a common aquatic ancestry, representing highly specialized forms of early water species. Millions of years ago, plants migrated onto land from aquatic environments, drawn by the promise of an abundant energy source in the sun, which would allow them to grow larger and to live longer. Once on dry land, plants underwent a series of adaptations that enabled them to optimize the process by which they harness energy. The first of these adaptations was a root system that anchored them into the ground and allowed them to constantly draw water and nutrients from the substrates around them. They also evolved a layer of skin, or epidermis, that prevented water loss, as well as pores in their leaves that enabled gas exchange. From there, the divergence of vascular plants into all the forms that occupy Earth's surface today was a process driven by further adaptation, specifically to climate and habitat.

Some land plants found their terrestrial homes unsuitable and eventually migrated back to aquatic environments. These plants that returned to water are extraordinarily diverse in their outward appearance, but as a group, they collectively differ from their landlubber cousins in basic ways, including in their leaf anatomy, in their rigidity (since they are buffered by water pressure, they do not require the stiff supportive structures known to land plants), and in their dependence on root structures, which are often reduced due to the abundance of water in their habitats. Many aquatic plants also have a reduced cuticle, the waxy skinlike covering that retains moisture and is important particularly in regions susceptible to drought and excessive exposure to ultraviolet radiation. As a result, aquatic species are dependent on either complete or partial submergence in water, and most cannot tolerate extended periods of time out of water.

Aquatic plants also may possess adaptations for growth in low-light conditions, since underwater environments typically receive less light than terrestrial environs. To improve light absorption in relatively dark environments, these species frequently possess special floating structures that lift them toward the water surface, where they can have greater access to sunlight. Certain of their leaves may lie flat to facilitate floating and sunlight absorption. These leaves often have a well-formed cuticle on their upper surfaces to protect them from water loss. Yet, for all these differences, aquatic species are fundamentally similar to land plants. For example, in relying on photosynthesis to produce energy, just like their counterparts on land, aquatic plants generate oxygen, which is important for the underwater environments they occupy. And, just as land plants provide habitat for land animals, so aquatic plants provide vital habitat for aquatic animals, including insects, invertebrates, and fish.

There are many different types of aquatic plants. Common hornworts, which are classified in the genus *Ceratophyllum*, are examples of aquatic plants that have mastered life in water. The most common species of this group is coontail, which is widespread in the temperate regions of North America, where it thrives in ponds, marshes, lakes, and streams. Other commonly encountered aquatic species include pondweeds, waterweeds, common mare's-tail, parrot feather, and quillworts, which are closely related to ferns. A new quillwort known as *Isoëtes eludens* was reported in 2007 by scientists working with the Royal Botanic Gardens, Kew. The small aquatic plant was found in the arid region of Namaqualand, Northern Cape,

South Africa, where it inhabited temporary pools of water, surviving dry phases in a state of dormancy. Its discovery is a testament to the current notion that only a portion of Earth's plants are known, that many species remain to be discovered and characterized.

Photosynthesis is an elegant process, in which plants cleanse carbon from the air. Carbon dioxide is absorbed through microscopic openings in leaves known as stomata, which are found specifically in the leaf cuticle. Stomata consist of a pore surrounded by guard cells, which control the extent to which the pore opens. These structures literally are like little mouths, inhaling carbon dioxide for the plant and absorbing or releasing water vapor. The leaves of plants are covered with stomata, so the shapes, sizes, and arrangements of leaves betray useful information about how a plant physically captures sunlight for photosynthesis. These factors also provide very valuable information about how a plant uses its leaves for activities such as defense and nutrient storage. With respect to the latter, leaves often are oriented with their flat surfaces angled such that they can efficiently collect sunlight during the day.

The trapping of sunlight is the work of the green pigment chlorophyll, which is found in chloroplasts, little photosynthetic engines within leaf cells. In the leaves, much of the carbon dioxide and water that a plant absorbs funnels into photosynthesis and is used to generate sugar for the production of cellulose, which accounts for 40 percent of the material found in the plant's cell walls. The walls are fundamental to providing rigidity to the plant, enabling it to withstand climatic factors such as wind and rain. The walls also provide yet another shield against the penetration into the plant interior of infectious organisms and potentially harmful rays of ultraviolet light. Plant cells also use photosynthesis for the production of amino acids, which are combined to form proteins, such as hormones that regulate plant growth and enzymes that catalyze photosynthetic reactions. Similarly, the production of fatty acids enables the synthesis of compounds that make up a plant's oils.

Many end products of photosynthesis not only support the survival of the plant but also represent valuable sources of nutrients for animals, including ourselves. But our relationship with plants is based on more than simply the nutrients they supply. If we skip backward through the evolutionary history of life on Earth, we find that all the different groups of organisms on the planet converge toward one another, becoming increasingly similar until they can no longer be distinguished and until ultimately we land on one very ancient group of organisms, the first life-forms on Earth. In fast-forward, we see life spreading from these first, single-celled forms, branching at first into two groups, then three, then four, until, after more than 3.5 billion years have passed, the tree of life has grown so many branches, each representing a group of organisms that diverged from their ancestors, that we find the lineages of mammal and plant split very, very long ago.

Despite the obvious differences in physical appearance and behavior, because both humans and plants are multicellular organisms we do share some fundamental similarities. One commonality can be seen in the microscopic makeup of our cells, which are eukaryotic in form, meaning that they are compartmentalized, with the compartments represented by organelles, membrane-bound machines that carry out functions such as DNA replication, energy production, and protein synthesis. All animals and plants are eukaryotes. In fact, Eukarya is one of three domains of life, into which all life-forms are grouped. Bacteria and Archaea are the other two domains, collectively representing single-celled organisms known as prokaryotes, being distinguished from the eukaryotes because their cells do not contain membrane-bound organelles. So, by cellular organization alone, we are already much more closely related to plants than we may realize.

There are many similarities in the cell molecules and organelles known to humans and plants. For example, plant chlorophyll is similar in structure to the pigment hemoglobin found in our blood cells, though hemoglobin's function is to carry oxygen, whereas chlorophyll captures sunlight. Plant and animal cells also contain many of the same organelles. However, the chloroplast is an organelle unique to plants and is responsible for energy production, a function carried out by mitochondria in our cells. Our mitochondria contain their own little genomes, consisting of a small number of genes. Chloroplasts have the same. Both organelles also are self-replicating, able to multiply in number in the absence of whole-cell replication. Other than the nucleus, which contains the majority of a cell's DNA, mitochondria and chloroplasts are the only organelles known to possess genetic material. They also are the only organelles to undergo self-replication. In the grand scheme of evolution, the unusual properties of mitochondria and chloroplasts suggest that these organelles may have been prokaryotes themselves at one time. Having invaded or having been engulfed by other cells, they presumably became organelles. This endosymbiotic theory remains one of the leading explanations for the evolution of these eukaryotic organelles.

There are a number of other distant comparisons that can be drawn in structure and function between the anatomical and cellular components of plants and animals. Similar to the xylem and phloem, which transport water, minerals, and other nutrients to the various parts of plants, we have blood vessels that serve to transport oxygen and nutrients to our organs. In plants, veins that diverge into the stems and leaves exchange molecules with those contained in the cells. The molecules absorbed into the vessels are then distributed to other parts of the plant, either for storage or for immediate use. Our blood vessels serve precisely the same function, with arteries diverging into our various tissues, branching into capillary beds, where nutrients are delivered to hungry cells and wastes are carried away through veins. Similar to the structure of our skin, the roots of plants are composed of layers of endodermal and epidermal cells. The basic layout, with an outer, protective epidermis, a middle cortex layer, and an inner endodermal layer hugging the vessels, is comparable to our outer epidermis, middle dermis, and inner subcutaneous layers. Of course, as plants such as trees age, their epidermal layers form one on top of the other, so that the older epidermal layers are grown over by a new layer each year. In contrast, our new epidermal cells are formed beneath our older epidermal layers, so that eventually the oldest cells are pushed up to the outermost skin surface, where they die and flake off. In trees, the encapsulation of epidermal layers through the process of secondary growth produces the characteristic growth rings that enable a dendrochronologist to estimate a tree's age.

Some plants are unusually animal-like. Venus flytraps, for example, respond to touch in very much the same way that we do. A flytrap can snap shut at an astonishing speed—within three-tenths of a second—when the tactile receptors, which lie on the edge of the trap, are stimulated. The behavior of this curious plant illustrates two other commonalities shared between plants and animals: nervous-system-like responses and carnivory. Venus flytraps snap shut through what has been described as a "nerve-like" biochemical mechanism. Neurons in animals transmit electrical impulses using shifts in ion concentrations within cells. These shifts cause a "depolarization," or activation, response, which enables the impulse to be sent along a string of neurons. The tactile-sensitive cells of Venus flytraps have ion channels that respond to changes in ion concentrations, with the result that the ion-generated signal triggers the trap to close when it "feels" an insect brush its receptors. Taken together, the shared cellular, anatomical, and physiological features of animals and plants may speak to more than simply coincidence. They suggest that although plants and humans are certainly very distantly related in the tree of life, similar natural law was at work in our evolution.

Carnivorous flytraps represent a somewhat darker side of plants. But insect-catching traps are a tool by which only a small subset of plants eat. In contrast, features such as thorns and poisons are mechanisms of defense potentially lethal to herbivores and other creatures that represent threats to a plant's survival. Thorns serve the function of physical deterrent to herbivores. Cacti possess the greatest arsenal of thorns. A light brush against some species can land a dozen tiny hairlike and excruciating spines in human skin. If they become embedded under the skin, they can give rise to infection. Thorns in other plants are equally menacing. The long, pointed, and branched spines of the honeylocust tree can create wounds painful enough to warn off any creature that incautiously consumes its sweet pods. *Gleditsia triacanthos*'s spines are so sharp that they can be used as sewing needles.

In plants, a major line of defense against herbivores and insects, as well as against plants competing for the same habitat, involves chemical warfare. The toxic chemicals plants produce belong generally to a group of substances known as secondary metabolites, since they are presumed to be unnecessary for the basic processes of photosynthesis, growth, and reproduction. The chemicals received this designation on the initial premise that plants can survive without them. Secondary metabolites, however, tend to give plants a competitive advantage in a habitat. In fact, they appear to endow plants with superpower-like abilities, enabling them to dominate their niches and lay claim to their own little sanctuary of soil.

In a phenomenon known as allelopathy, the growth of one plant species is inhibited by secondary metabolites produced and released into the surrounding environment by another species. Allelopathy, which means "reciprocal suffering," is complex and requires just the right mixture of chemicals to work. The camphor tree, for example, a species native to Asia and cultivated in southern regions of the United States, produces large quantities of camphor, which in conjunction with certain other compounds can inhibit the growth of seedlings of other plant species that try to take root near the tree. Allelopathy explains why tomatoes are unable to grow near walnut trees and why certain weeds are unable to grow in the presence of sunflowers. It also underlies the ability of certain invasive plants to take over the habitats of native species.

Secondary metabolites often have medicinal actions in humans, and when present in leaves or other plant parts, they frequently impart a disagreeable, bitter taste. Plants containing these substances are sometimes ingested intentionally by animals. Chimpanzees living in the wild in Kibale National Park in Uganda, for example, eat more than 160 different types of plants, about 35 of which are used in various systems of traditional medicine by humans. The chimpanzees consume the medicinal plants despite the fact that the plants themselves may be of very little nutritional value. It is suspected that the animals, when seeking relief from ailments like skin irritation or intestinal parasite infection, are instinctively driven to eat plants that are actively producing secondary metabolites.

Many of the discoveries concerning allelopathy and secondary metabolites occurred in the twentieth century and thus are recent additions to the study of plants. The field of botany as we know it today, encompassing investigation of the biochemical properties of plants, the diseases that affect them, and their classification, is considered to have been established in the early 1500s by three Germans, one of whom was Otto Brunfels.

Brunfels developed an appreciation for plants within the sanctity of a monastery. Between 1530 and 1536, he published a work titled *Living Pictures of Herbs* (*Herbarium vivae eicones*). It contained some of the most accurate descriptions and drawings of herbs and other plants up to that time. The illustrations were created from direct observations of the plants as they existed in nature. Brunfels's work was significant not only because of its scope and illustrations but also because it was original. Nearly all other works on plants composed in Europe in the Middle Ages were copied, including the *Juliana Anicia Codex*, one of the oldest illustrated botanical works known. The *Codex* was believed to have been originally written by Greek physician Pedanius Dioscorides, sometime in the first century CE, and titled *De Materia Medica*. The work was believed to have been later illustrated by a Byzantine artist in 512 for the daughter of Roman emperor Olybrius. Her name was Juliana Anicia.

The *Anicia Codex* contained magnificent drawings of the plants that Dioscorides described, but the illustrations themselves were an anomaly in the otherwise disembodied art characteristic of the Byzantine Empire. This raised speculation among historians and led to an investigation of the drawings. Researchers found that the illustrations actually dated to the second century CE and were the work of Greek pharmacologist Crateus. Thus, the drawings in the *Anicia Codex* are reproductions. Their Greek origin was further supported by the fact that the plants illustrated, which included wormwood, fennel, rose, black nightshade, and winter cherry, were native to Greece. Dioscorides's work, however, has had a significant influence on modern botany. A number of scientists and botanists have highlighted the importance of his written descriptions in attempts to discover new plant compounds for modern drug development.

A contemporary of Brunfels, and the second contributor to the establishment of botany, was German botanist Hieronymus Bock, who published *Plant Book (Kreuterbuck)* in 1539. Though it was not illustrated, the work provided details on the physical characteristics and uses of each plant he knew. Bock's work stands out because he was the first to use a classification system.

The third individual to help establish the foundations of modern botany was German physician Leonhard Fuchs. Fuchs published *Notable Commentaries on the History of Plants (De Historia Stirpium Commentarii Insignes)* in 1542, which included illustrations and comprehensive descriptions of some 400 plants. Although Fuchs's work is the first to include a plant from the New World, the chile pepper, he did not recognize it as a species not native to Europe. In fact, it appears that he was unaware of *Capsicum*'s origins altogether.

New World plants were described in comprehensive fashion for the first time by Spanish botanist and physician Nicolás Monardes. He received plants brought back to Seville following expeditions by Spanish explorers and transplanted the various specimens into his garden. This gave him time to study the plants in great detail, identifying their distinguishing characteristics and attempting to understand their medicinal applications. To study the latter, Monardes tested plant extracts on animals. His most famous work, a compilation of several volumes, was completed in 1574. One part of this work was called *Dos Libros*, released in 1569, which contained a description of tobacco. Prior to this publication, Monardes had been skeptical of the medicinal value of New World plants, at one point even denouncing them as inferior to the plants of Spain. But he later humbled his national pride. He came to believe so strongly in the medicinal properties of New World plants, in fact, that he persuaded the Spanish king, Phillip II, to have the explorers collect more extensive information on the plants and use the medicines derived from them as treatments.

Garcia de Orta, a physician in Portugal, took it upon himself to expand European knowledge of plants by writing about medicinal plants of the East. De Orta, similar to Monardes, never traveled for his studies; instead, in keeping with the trend of his time, he maintained gardens filled with plants brought back by explorers who had painstakingly collected each specimen. He published works describing plants from India, China, and Persia. His descriptions of a number of plants were based largely on the products and accounts given to him by merchants traveling from those lands.

The European world became acquainted with the medicinal plants of Mexico in the middle of the sixteenth century through the writings of an Aztec. More than 180 plants used as medicines by the Aztecs were described and illustrated in a single work, known as the "Badianus Manuscript." In contrast to European works on plants, which often began with plant etymology and were organized in alphabetical order, the Badianus Manuscript was organized by disease. The author, Martín de la Cruz, supposedly composed the work in an effort to demonstrate to Spain's Charles V, who was distracted

Jean Nicot presenting the tobacco plant to Queen Catherine de Médicis in 1561. (Photo credit: Library of Congress, Washington, DC; copyright by Levy Bros.)

by his quest for conquests of peoples all over Europe and the Americas, the knowledge of the Aztec people and the importance of their continued education at the Colegio de la Santa Cruz in Tlatelolco, which was running out of money and needed the king's support. The manuscript was the first from the Americas and is devoid of European philosophy. Also, since none of the plants had Latin names, their Aztec names were maintained in the text that was translated into Latin by Joannes Badianus. Tobacco, for example, was known as *picietl* to the Aztecs—not *Nicotiana*, a name given to the plant in the eighteenth century by Carolus Linnaeus in honor of the French ambassador to Portugal, Jean Nicot, who introduced the plant to the queen of France, Catherine de Médicis, in the sixteenth century.

It was around the time that Monardes was prodding Phillip II to send someone to collect information on the native plants of the New World that Spanish royal physician Francisco Hernandez was shipped over by his royal highness to see about the native medicines.

Hernandez went to the Huaxtepec gardens, which had flourished during Moctezuma I's reign as Aztec emperor in the fifteenth century. The gardens were filled with medicinal plants, and Hernandez met with local Aztec herbalists to learn about the plants and their medicinal uses, eventually returning to Spain seven years later with an abundance of information.

The discovery of the biological properties of plants was a much more gradual process than the identification and illustration of new species collected by the dozens by explorers. A series of observations made in the 1700s, however, eventually culminated in major breakthroughs in the basic understanding of how plants work. The most significant discovery was the realization that the leaves of plants are capable of absorbing sunlight and air from their surrounding environment. The theory for this was first put forth by English physiologist and chemist Stephen Hales in the 1720s. He suspected that plants used the air as a source of nutrients and the light from the sun to fulfill some other function. With the power of a microscope, an instrument that was invented only little more than a century earlier, Hales observed plants perspiring. In a brilliant study, he decided to measure the amount of water being excreted during this process and to compare this measurement to the amount of water being taken up by a plant's roots. He found that the roots absorbed lesser quantities of water at night, which indicated that some process that occurred during the day within plants was requiring them to absorb water.

Hales's hypothesis that plants used nutrients from the air was supported by the work of English scientist Joseph Priestley, who demonstrated that plants had the ability to "restore air" (release oxygen). Jan Ingenhousz expanded upon this by demonstrating that plants are capable of restoring oxygen to the air only in the sunlight. He also theorized that animals were the ones that produced carbon dioxide, which plants then utilized and returned to the atmosphere as oxygen. Though Ingenhousz did not recognize it by the name "photosynthesis," he is often credited with its discovery. Hales's discovery that plants absorbed more water during the day than at night fit perfectly with Ingenhousz's work, since photosynthesis requires water, and when the sun goes down, a plant's need for water decreases.

In the century that followed, botany turned to understanding how plants reproduced, fueled by the work of Carolus Linnaeus, who not only devised a system of nomenclature for genera and species for animals and plants but also hypothesized a system of reproduction for plants whereby the stamens and styles represented male and female parts. His insights on sexual reproduction in plants were published in 1735 in *The System of Nature* (*Systema Naturae*), a work considered progressive for its scientific detail and provocative for its use of human social and sexual constructs to describe plant reproduction. Most notably, Linnaeus invoked the concepts of marriage and used terms like *wife* and *husband* to describe the various sexual relationships in flowers. He placed husbands and wives in the same bed, in separate beds, and even in separate houses, depending on a plant's arrangement of reproductive parts.

A year after *The System of Nature* appeared, Linnaeus published his principles for botanical nomenclature in *The Foundations of Botany* (*Fundamenta Botanica*). Both of Linnaeus's works captured the imaginations of succeeding generations of scientists, including English botanist and physician Erasmus Darwin, who explored in great detail the various facets of plants, including their beauty, their evolution, and their medicinal uses.

Ecology, biogeography, genetics, and biochemistry all are important elements of botany today. Much of the knowledge that exists in the modern era, however, originated long before Brunfels, Bock, and Fuchs, with the peoples of ancient civilizations, who relied on simple intuition or trial and error when deciphering poisonous from nonpoisonous plants and on the basic senses—sight, smell, touch— when observing them. Those of us alive today are not dependent on such visceral experience and connection to nature to understand the mechanics of plant function. We can learn about it in books. In distancing ourselves from the outside world, however, we have lost touch with not only the fundamental reasons for appreciating plants but also the consequences that our activities can have on their well-being.

Our desire to connect with nature is in itself an interesting subject, one that encompasses both an instinctual drive to interact with other forms of life and the various aspects of modern society that have smothered this behavior. Learning to reconnect with nature— and delving into the complexities of the human interaction with the environment—touches on the meaning of humanness, which goes beyond chemicals and cells. Rarely are drug discovery and human-ness mentioned in the same sentence, even though drugs affect our bodies and minds, and our bodies affect the actions of drugs. The combination of these factors influences our behaviors. But how can we better appreciate the interaction between ourselves and these substances that are meant to heal us and that drive a multibillion-dollar industry? One way, and hopefully one that is effective, is to place plants, drugs, and ourselves into the context of life on Earth. We can learn a lot about human nature from our direct interactions with plants—gardening, commercial agriculture, logging—and from understanding the instincts that underlie our desire to explore the natural world.

Manifestations of the human drive to explore, such as expeditions by botanists and drug prospectors to remote lands, serve as vehi-cles to help us see more clearly how humanity, drug discovery, and plants are intertwined. Exploration rekindles our curiosity of the animate and inanimate of the natural world and is associated with an instinctual drive known as biophilia, our love of nature and all the life it contains. To better understand how these elements affect our relationship with plants, we will embark on an expedition of our own, into the basic human behaviors that underlie the adventure and discovery of exploration and plant hunting.

3

The Biophilia Factor

Leaves and seeds of the neem tree (*Azadirachta indica*), native to India.

IN THE 1980S BIOLOGIST E. O. WILSON proposed that humans possess an innate love for all forms of life. He described this instinctive human attraction to nature as biophilia, and this perhaps explains why many people enjoy activities such as hiking and bird watching and why we support programs for environmental conservation and preservation. It probably also explains why many of us perceive natural forms as beautiful and intriguing. As scientists discover new information about these forms, we become even more fascinated by them.

Wilson's own fascination with nature, much like that of many other people's love of the outdoors, began at a young age, nurtured by the freedom of time for exploration and adventure. The greater part of that formative period of his life was spent in the Florida panhandle and near the southern, coastal region of Alabama, where he ventured on numerous occasions into the woods and swamplands in search of all sorts of animals, from snakes and frogs to birds and insects. As an adult, his investigations of the natural world became focused on insects and other small creatures, which he later described as "composing the foundation of our ecosystems." One of the defining factors of his life, which is evident in his work and in his easy, calm manner, is that above all else his love and appreciation of nature remained a constant and undeniable force.

Wilson's research centered on the study of ants, a field known as myrmecology. Although the subjects of his research were tiny, the impacts of his observations of the inner workings of ant societies and his theories of the human relationship with the natural world were exceptionally broad and far-reaching. Wilson spent a great deal of time thinking about complex questions in biology, especially those pertaining to the behavior of social beings, from insects to humans. One of his lesser-known works was *Biophilia* (1984), which described in succinct and simple terms a new hypothesis about humans and nature. Wilson landed on his conceptualization of biophilia by way of observation. He compiled information about his own interactions with nature and reflected on his passion for wanting to learn more about Earth's living systems. He couldn't help but notice that this same interest in the natural world was, more or less, common to all humans.

Biophilia is a concept that was explored initially in the mid-1960s by Erich Fromm. Fromm was a psychoanalyst, and to him biophilia meant "a love for humanity and nature." As Wilson delved deeper into attempting to interpret the human relationship with plants, animals, and the environment, he too came to believe that all humans inherently love nature. Wilson modified Fromm's definition, describing biophilia as "the connections that human beings subconsciously seek with the rest of life." When *Biophilia* was published, there was virtually no hard evidence in place to uphold the ideas it presented. Still, Wilson stood behind his new hypothesis, refusing like a rock to budge on any argument against it. His stubbornness emanated from quiet confidence and skepticism, which kept him asking questions about the world. His theory became extremely difficult to deny when he eloquently stated that "the biophilic tendency is nevertheless so clearly evinced in daily life and widely distributed as to deserve serious attention."

Because biophilia relates to everyone and appears to be indifferent to culture, Wilson suspected that it must be, at least to some extent, and perhaps even to a very large extent, a behavior programmed into the human genetic code. This notion, as well as qualitative evidence to support the actual existence of biophilia, was presented in detail in *The Biophilia Hypothesis*, which Wilson co-edited with social ecologist Stephen Kellert. The work summarized the role of biophilia in establishing a conservation ethic and provided explanations for a biological foundation of the behavior. It also described the influence of our relationship with nature based on the human experience— from biophobias to symbolism to culture and politics.

As Wilson explained, our biophilic drive manifests in a variety of ways and likely influences many of our decisions, including where we choose to live, where we choose to vacation, and whether we choose to own pets. In the late 1990s, interest in the biophilia concept lost momentum in biology, having been considered more pertinent to the study of human behavior. Biologists and ecologists were more concerned with continuing their efforts to identify new species, to preserve nature, and to protect endangered species and habitats before it was too late. They were less interested in sifting through the quagmire of the human genome, in search of a genetic explanation for our biophilic drive.

For a while, a single "biophilia gene" was believed to exist. But as new information became available about the complex interplay between our genes and our biological traits, the one-biophilia-gene hypothesis expanded, so that now our love for nature is suspected to be linked to multiple genes, perhaps hundreds or even thousands. It is a dizzying thought that so many individual genes can contribute collectively to this singular behavior. But for scientists, this complexity is rapidly trivialized by the more troubling notion that in modern humans some of these biophilia-contributing genes may no longer be active, which means that they would be exceptionally difficult to identify in the first place. Beyond this lies the problem of associating presumed biophilia genes with actual biophilia behavior, a far from simple task. All of these difficulties have challenged efforts to compile significant molecular evidence in favor of biophilia. And today, as leaders struggle to agree on global climate change policy, as urban sprawl consumes what remains of the world's wilderness, and as plastics swirl into continent-sized patches in the oceans, it has become more important than ever to identify and reinvigorate the biological components of biophilia.

Our daily lives mask the reality of our dependency on the natural world. Our adaptation to technology has diluted our biologically programmed biophilic drive. Even Wilson acknowledged that our diminished exposure to the environmental stimuli necessary to achieve full expression of biophilic genes has likely weakened our desire to connect with the natural world. Still, he was optimistic that this innate human behavior can be reawakened.

The existence and goals of conservation programs that incorporate educational elements into public awareness have captured our attention. They have helped us understand how our actions impact the health of ecosystems and have sparked enthusiasm among our youngest generations to explore nature and to protect the environment. Rekindling our appreciation for nature can be done in many different ways, whether through gardening, organic farming, learning about conservation, exploring the outdoors, or studying it scientifically. Each of these steers us down the path toward environmental awareness and stewardship and helps us establish a conservation ethic, as individuals and as societies. Of all these endeavors, however, learning about Earth's biological life and the way it functions fulfills perhaps the most vital role of all. Bioliteracy helps us take meaningful steps toward rejuvenating our individual biophilic drives. In our increasingly technological world, our knowledge and understanding of biological forms other than ourselves are fundamental to the process of generating human concern for nature. Such knowledge has the potential to defeat monumental environmental challenges, including global warming and pollution, which are currently wreaking havoc with life on Earth. Every moment in our lives presents an opportunity for us to learn something new about our relationship with nature. One of the most basic elements defining this relationship is cooperation, which is illustrated elegantly by plants and the organisms with which they associate.

Just as we benefit from seeking connection with nature, so too are benefits experienced by other life-forms that seek to connect with one another. Plants are great subjects for exploring connections between living entities in the natural world. They partake in a variety of different and often complex interactions with other organisms. For example, plants do not produce chemicals only to defend against predators or for purposes of allelopathy. They produce them also to attract comrades—insects, fungi, and other living creatures—that aid them in their struggle for survival.

A well-characterized example of symbiotic relationships between plants and fungi are mycorrhizae, in which fungal tentacles (technically, hyphae, which are long filaments that look like threads) grow and infiltrate a plant's roots. The tentacles spread out into the surrounding soil and absorb nutrients, which are then shared with the plant. The plant cannot obtain these extra nutrients on its own, and so in the presence of an infiltrating fungus it is able to grow larger and more quickly, gaining an advantage over the plants around it. The fungus benefits too, drawing small amounts of sugars and other carbohydrates needed for its growth from the plant.

Cooperation and biological attraction in nature are also seen in the form of pollination, whether by insects, birds, or other animals. Plants that conspire in this somewhat provocative process with insects are referred to as entomophilous, and this form of pollination has a long natural history. According to scientists investigating the prehistoric evolution of insects known as scorpionflies, pollination actually predates the existence of flowering plants. Scorpionfly fossils have revealed that these insects possessed specially adapted heads and mouthparts, which are believed to have evolved specifically for the pollination of gymnosperms, conifers and cycads with seeds described as "naked," or not contained within an ovary but that occur in cones. Gymnosperms appeared on Earth before flowering plants, and scorpionfly pollination of gymnosperms might have occurred perhaps as many as 62 million years before the appearance of flowering plants (the latter appeared an estimated 140 million to 180 million years ago). At the root of pollination is the ability of a plant to make itself attractive to its servicer. The production of nectar, of large or brightly colored flowers, or of particular scents are just some of the mechanisms employed by entomophilous plants to seduce specific types of insects, whether bees, butterflies, or beetles. In return for their efforts, pollinators are rewarded with various highly sought after prizes, such as oils, sugars, or resins.

In some cases, one type of animal, such as bees or hummingbirds, serves as the entity responsible for the propagation of a single species of flowering plant. Such a relationship is the result of thousands of years of coevolution and specialization, and it is most apparent in particular species of orchids, which are visited by specially adapted pollinators. Some types of orchids are particularly conniving when it comes to ensuring their sexual success. They engage in food deception, in which they offer false hopes of food that they ultimately do not provide to the pollinator, and in sexual deception, in which they produce a sex pheromone that attracts a male insect and encourages it to pseudocopulate with the flower. These forms of orchid deception are individualized, involving one animal species and one species of orchid. The orchid *Chiloglottis trapeziformis,* found in Australia and New Zealand, uses sexual deception to attract male thynnine wasps (*Neozeleboria cryptoides*). As the wasp attempts to mate with the orchid, he picks up pollen. Disappointed with the lack of reciprocation on the part of the flower, he buzzes away to another flower of the same orchid species, where he brushes up against the female flower parts with his pollen-coated back as he again tries to mate with the plant.

Orchids are exceptional in their relationships with pollinators. Most other plants have evolved to be dependent on relatively promiscuous insects, which visit multiple, different plant species. These insects are commonly referred to as generalized pollinators. Plants visited by promiscuous insects must expend more energy producing pollen in order to ensure their propagation. In contrast, plants engaged in an essentially monogamous pollinator relationship, with a high rate of successful pollination, can conserve far more energy, which can be channeled into various activities, including growth or defense.

Pollination forms only one stage of the reproductive cycle of flowering plants. Once a fertilized ovule has developed into a seed, the surrounding ovary tissue enlarges to form a fruit. The dispersal of seeds from fruit is the next step in propagation. A plant may depend on elements such as wind, gravity, or water to spread its seeds, or it may depend on an animal. One of the reasons why plants are among the most successful species on Earth is that they take no shame in exploiting the physiology of animals. They essentially hypnotize pollinators, luring them with irresistible colors and fragrances, and they use coercive forces on animals, enticing fruit eaters with shiny, sweet bounty and then waiting for them and their digestive systems to dispense with the seeds in new locales. Humans, of course, are just as susceptible to enticement by the succulent sight and taste of fruit. Watching people in an orchard blanketed by the musty scent of ripe fruit conjures up notions of a saturnalia festival. The pleasure and indulgence of humans and other animals, however, are only temporary. For the plant it is everlasting.

The reliance on animals for seed dispersal developed over the course of millennia, with flowering plants ultimately finding an advantage in the gastrointestinal tracts, fur, and food-hoarding behavior of animals. Those plants with distribution that is dependent on animal digestion typically have seeds designed for passage through the harsh environment of an animal's stomach and intestines. Some seeds even have an extra-tough outer coating because they are digested by a species with exceptionally caustic stomach juices. In other cases, dispersal is accomplished simply by the presence of some adaptive seed feature such as burrs, which allow the seeds to hitch a ride on an unsuspecting animal's fur until they eventually lose their grip in some distant place. Animals such as squirrels and certain birds, including acorn woodpeckers and jays, bury seeds in caches, which can be accessed when food resources become scarce. In the meantime, however, if the seeds are buried in a site amenable to growth, they may fall under the persuasion of basic biological forces and take root. Regardless of how it is accomplished, animal seed dispersal is beneficial in that it enables plants to be distributed over wide ranges since the animals wander away from the parent plant, sometimes to distances of tens of miles.

Certain species of plants have enjoyed other types of intimate connections with the life around them. For instance, acacia trees and stinging *Pseudomyrmex* ants enjoy a mutually beneficial relationship. These two vastly different forms can be found residing together in tropical regions of the world. Acacia trees depend on the truculent nature of the ants to defend against hungry grazers, particularly various types of insects. The ants sometimes even mow down other species of plants that encroach on the acacia's territory. In return, the ants feed on protein-rich parts of the acacias and live safely within the tree's enlarged thorns, which are hollow and provide cozy nesting sites.

Mutualism is driven by the fact that each participant reaps some benefit from the relationship. For that reason, it is not necessarily the same as biophilia. A significant aspect of the biophilia hypothesis is recognizing the right of Earth's many different forms of life to exist, even though to us they may serve no immediate or apparent use. We live within a larger ecological community, and though we may reside at the top of the community pyramid, we are supported by a sprawling, complex system of life. Within this system we are bound by dependencies and interactions, many of which remain incompletely understood scientifically but are respected as being fundamentally important ecologically.

Many facets of the community ecology principle are observable in nature, and one of the most obvious is the natural harmony that exists between various and diverse life-forms in a confined area. In the tropical rain forest, for example, the richest place on Earth in terms of biodiversity—literally thousands upon thousands of unique creatures—inevitably there are populations that compete with one another. Intermingled amid the rivalries and contests, however, are efforts of cooperation and at least a semblance of respect, which are vital in maintaining the balance of the greater whole of life in the community. The loss of a single species, or even just a decrease in the productivity or population of a species, is enough to throw the balance of the entire community askew.

Modern human civilizations are becoming increasingly notorious for pushing ecosystems off balance, with certain activities gradually endangering species and wiping out biodiversity. We cut down trees in our forests without noticing the bird nests that careen to the ground or the families of opossums, squirrels, or mice living within the trunks that flee in panic at the sound of the chainsaw. There are rhinoceros and tigers and whales that stand defenseless against poachers' weapons, plants that are choked out by invasive species, and human inventions such as the two-stroke engines found in many ATVs, snowmobiles, and jet skis that pollute our air and water. At the top of the senseless-destruction-of-nature scale are warfare and nuclear warfare, which obliterate not only the existence of humanity— human lives, culture, and history—but also all creatures and ecosystems for miles around.

Despite what we tell ourselves about our ethical or religious integrity, our many daily activities are slowly and silently ravaging the environment, and it can be argued that these activities, though a part of our modern life routines, qualify as senseless destruction. Many people do not struggle with justifying their need to drive automobiles or their persistent consumerism, since these activities are perceived as necessary economically. The economy has almost always reigned victorious over environmental interests, with the result that the well-being of the environment has been compromised, not only at the expense of its biological function but also at the expense of cultivating human compassion for and interest in nature. The latter elements are where biophilia comes into the picture. Our innate attraction to the world beyond our windows is superbly demonstrated by ancient and modern explorations into nature, many of which were conducted by naturalists and botanists who endured every physical, mental, and financial assault imaginable in their quests to discover new species.

Throughout history, humans have journeyed into nature to gather plants. This was performed initially with the intention of collecting plants for food, such as by peoples who subsisted by hunting and gathering. In most parts of the world, this mode of plant collection was replaced by the cultivation of domesticated crops, such as wheat and barley. Later, naturalists began to travel far and wide for the sole purpose of discovering new and exotic species, an activity that certainly was not limited to plants. Naturalists snatched up specimens of whatever living creatures were unknown to them, including insects and peculiar mammals. The various things they brought back with them were identified, cataloged, preserved, and often made part of someone's private collection. In some cases, they were put on display in museums to share with the public. In the case of plants, coveted seeds were stashed at well-funded botanical gardens and nurseries, sold at high cost to the public, or given to only a few wealthy individuals.

Cultivating trees and other plants with elaborate, unusual, and beautiful flowers is an obsession for many. Among the most popular plants on which humans first doted were fruit trees, fig trees in particular. Figs uncovered from an archaeological site in Jericho were estimated to be 11,400 years old and are believed to have been grown specifically for food. In the greater timeline of agricultural history, the cultivation of fig trees took place perhaps as many as 1,000 years before the first domesticated crops, namely wheat and barley, appeared.

Since the time of the ancient Greeks, new plants generally have been collected with the intention of cultivating them. In Greece, plant cultivation began long before the rise of the ancient civilization around 1200 BCE. But rather than individuals journeying away from Greece to find specimens, many plants for food and medicinal use were brought to the land along established trade routes. The primary routes led to and from the Near East, which encompassed Southwest Asia, including the Arabian countries, and northeast Africa. Several of Europe's early domesticated crops came from the Near East, having been first introduced by way of Greece. One of the earliest known agricultural plants to appear along the Mediterranean was barley, arriving sometime between 6000 and 5000 BCE and being farmed with great success.

Although plants had been collected for various purposes for thousands of years by the ancient Greeks, and by other peoples worldwide, the first written record of an expedition meant specifically to gather exotic plants is Egyptian, dating back to 1495 BCE. The story relates details of Queen Hatshepsut's request that botanists travel to Somalia to collect incense trees. Many centuries later, in the 1600s, there was an upsurge in plant-hunting expeditions. These were conducted primarily by Europeans, particularly the British, and were driven by gardening, especially among the aristocracy, as well as by basic scientific interest and the desire to identify new plants that could be used in the practice of medicine.

Among the better-known plant hunters of the seventeenth century are the John Tradescants—the Elder (c. 1570–1638) and the

Younger (1608–1662)—both of whom made important and numerous contributions to English botany. While working for the first Earl of Salisbury, Robert Cecil, the elder Tradescant was sent to France and the Low Countries of Europe in 1610 and 1611. On those trips, he collected plants to add to the garden of his benefactor. He also traveled to Russia, in 1618, in search of new and unusual plants, and he later visited Algiers and Paris with his new patron, George Villiers, first Duke of Buckingham. Tradescant ended his career in Surrey, working as Keeper of his Majesty's Gardens, Vines, and Silkworms at Oatlands Palace.

Tradescant was known not only for his skills in plant cultivation but also for the impressive collection of specimens of all sorts—plants, animals, and various "curiosities"—that he brought back with him from his travels. The younger Tradescant also traveled to collect plants, venturing on three separate occasions to Virginia to gather North American species. He assumed his father's position at Oatlands Palace following the elder's death in 1638. Together, both Tradescants introduced a variety of new plants to England, including Virginia spiderwort, swamp cypress, the American cowslip, Michaelmas daisies, and the tulip tree.

Though the Tradescants' contributions to botany were great, the man who bears the title "father of modern plant hunting" is Englishman Sir Joseph Banks (1743–1820). Banks was born in London, the son of a wealthy, privileged landowner. The family name and wealth came in handy for Banks, essentially paying his way into Christ Church at Oxford University. In 1764 Banks inherited such a large sum of money after his father's death that he instantly joined the wealthiest ranks of men in Great Britain. There were so many things Banks could have done with his money. But he loved botany, and he chose to follow his passion, to travel to places around the globe in search of rare and exotic plants. In 1766 his first opportunity arrived, serving as naturalist aboard the HMS *Niger*, which sailed along the coastlines of Newfoundland and Labrador. He collected numerous plant specimens during this voyage, all of which he named and characterized.

Two years after his trip on the *Niger*, Banks learned that Captain James Cook would be leading an around-the-world voyage aboard a ship named *Endeavour*. There was enormous potential for studying plants at the various destinations en route, and Banks couldn't resist. He bought his way on board and paid the way for his nine crew members, five of whom assisted him in collecting, cataloging, and drawing specimens and four of whom acted as servants. The *Endeavour* sailed from England to the island of Madeira, then to Rio de Janeiro and Tierra del Fuego. After rounding the southern tip of South America, the ship proceeded west to Tahiti, then circled New Zealand, skirted the eastern edge of Australia, and brushed along the southern edge of the islands of Indonesia. After stops at Cape Town and St. Helena Island, the ship returned to England by way of cutting a sweeping arc through the North Atlantic Ocean. For Banks the three-year voyage of the *Endeavour* was one marked by surprise, danger, and disappointment, most of which stemmed from interactions with locals in the places they stopped. In Rio de Janeiro, for example, Banks was forbidden to paddle ashore by the viceroy, Don Antonio de Moura, who distrusted the reason for their stopover there. Banks was forced to sneak to land, to study and collect specimens of the native plants clandestinely during the day, returning to the *Endeavour* in the darkness of night. At Tierra del Fuego, Banks and his crew again paddled ashore. This time, though, they met with no political resistance to their activities, but they did meet with a menacing snowstorm. Hypothermia claimed the lives of two of his servants. When the ship arrived at Tahiti, Banks again found little time to track down new plants—he spent most of his time there dissipating potential outbreaks of fighting between the locals and the ship's crew.

Following Banks's return to England, he set to describing the many specimens that he collected despite the distractions. In the end, he cataloged a total of 110 new genera and 1,300 new species. He later served as the scientific adviser on plant life for Kew's Royal Botanic Gardens. The position was created for him in 1772 by King George III, who was captivated by Banks's work and befriended the adventurous botanist. Among the species that Banks discovered during his exploits were the everlasting strawflower from Australia, the chaura from South America, a kowhai legume from New Zealand, and a tree that came to be known as coastal banksia from Australia.

In 1772 Francis Masson, who was then an undergardener at Kew, set sail aboard the HMS *Resolution*, captained by the ever-faithful

Cook. Banks was originally scheduled to sail on the *Resolution*, but after finding the accommodations unsuitable, he recommended that Masson go in his place. The ship was destined for Cape Town. Despite having been supplanted from his comfortable life in England, Masson quickly adapted to his newfound role as a plant hunter. Between 1772 and 1773, he made three extensive journeys into the interior of South Africa, exploring different regions on each trip. His journeys took him to the Bokkeveld, Roggeveld, and Hottentots Hollands Mountains, and all points between. On these expeditions, he was accompanied, sometimes much to his annoyance, by Swedish naturalist Carl Peter Thunberg. Although Masson often found himself at odds with Thunberg, the two proved to be an effectual team. Masson returned to his motherland in 1775, having discovered hundreds of species previously unknown to Europeans, including the bird-of-paradise flower.

Back at Kew, it was clear that Masson had developed the itch for adventure. He longed to travel once again, in search of plants new to the then limited scholarship of English botany. During the years 1776 to 1785, Masson ventured to the islands in the North Atlantic, including the Madeira Islands, the Azores, and the Canary Islands, as well as to the West Indies and Tangier. In 1786 he returned to Cape Town, remaining there for the next nine years. In these later travels through southern Africa, he discovered the giant white arum lily. Other plants that Masson brought to the attention of European gardeners included the king protea, the belladonna lily, the perennial dusty miller, and the white trillium. He found the latter species on a trip to North America, where he died in 1805.

Another famous plant hunter from the Kew establishment was Sir Joseph Dalton Hooker. Hooker studied botany and medicine at the University of Glasgow. His first adventure began in 1839, with an invitation from Sir James Clark Ross to join the crew of the HMS *Erebus*, which was bound for the Antarctic. At each stop that the ship made to restock supplies, Hooker explored the local plants and collected specimens. In addition to the cold, dark dreariness of the Antarctic, the voyage took the young man to many botanically rich lands, including Madeira, the Cape of Good Hope, Desolation Island, Van Dieman's Land (Tasmania), New Zealand, and Hermite Island. Hooker's journey lasted three years. Soon after his return, he began arranging a trip to a different region of the world, India, which would enable him to compare the plants that grew there to those that grew in the cold climate of the Antarctic. He was most interested in visiting the Indian state of Sikkim, in the eastern Himalayas. In 1847 he finally set sail, a passenger aboard the *Sidon*, which took him to Alexandria, Egypt, where he boarded the *Moozuffer*, an Indian steamship bound for Calcutta (Kolkata). After arriving there, he proceeded to travel by foot to Darjeeling. His quest for the discovery of exotic plants had now begun in earnest.

While he waited to obtain permission from the local government to explore the depths of the mountains, Hooker managed to content himself with probing about the relatively limited confines of Darjeeling. His efforts landed him on the discovery of the magnificent *Rhododendron grande*, the yellow-flowered *Rhododendron falconeri*, and Campbell's magnolia. Eventually, after painstaking arguments with the local leaders, it was by way of military threat by the British that Hooker received permission to advance into the Himalayas. His journey through these rugged mountains would prove among the most fruitful expeditions in history to be conducted in the name of natural science. New species of insects, birds, and plants were among the many forms of life that Hooker and his team of collectors encountered. A host of new species of rhododendron and several species of wildflowers, including *Primula capitata* and *Primula sikkimensis*, were included among his finds. In the years after the conclusion of his Indian travels, Hooker visited Lebanon, Morocco, and North America. He collected a variety of seeds and specimens for Kew. He also became an important champion of Charles Darwin's theory of evolution, which he found especially applicable to plants.

Other notable plant hunters included Scottish botanists David Douglas and Robert Fortune. Douglas is perhaps best known for having brought the North American Douglas fir to Scone Palace. Fortune, who went on multiple expeditions to Asia, is noted for his introduction to European gardening of the Japanese cedar, green-stem forsythia, the Chinese rhododendron, winter jasmine, and another type of deciduous shrub known as *Weigela florida*. While Douglas endured his own share of difficulties, Robert Fortune's experiences reveal the extremes that obsessed plant hunters would

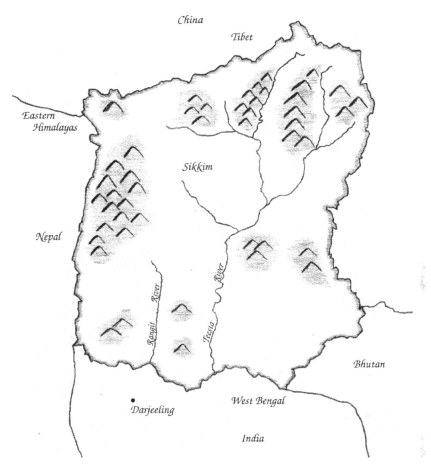

Sir Joseph Dalton Hooker discovered many different types of plants while in the Indian state of Sikkim, which lies in the eastern Himalayas.

go to in order to obtain specimens, on which not only their incomes depended, but their reputations as well.

Under the auspices of the Horticultural Society of London (now the Royal Horticultural Society), Fortune embarked in 1843 on an expedition to China, which had become recognized by English botanical connoisseurs as an important, unexplored plant frontier. Poorly equipped, in terms of finances and weapons to defend himself, Fortune endured robbery and a variety of assaults, including pirate attacks. He was assaulted by locals in most of the towns that he visited in China, and the persistent hostility toward outsiders led to his attempt at dressing like a local to avoid recognition. He found some inconspicuous clothes and assumed a Chinese character, even mimicking the native hairstyle. His efforts paid off. When Fortune returned to England in 1846, he had transferred to his homeland a variety of live plants, contained in Wardian cases (similar to modern aquariums), and trunks full of seeds and dried specimens.

Fortune visited China on two other occasions and made trips to North America and Japan. His practice of adopting a local costume to blend in played a vital role in his later trips to China, enabling him to discover tea plants in the city of Suzhou and in the nearby Zhejiang Province. One of his most significant contributions to British culture was his establishment in the mid-1800s of tea farms in the foothills of the Himalayas, having delivered to India not only live tea plants but also Chinese tea-growing experts.

One of the most recognizable plants known today—the pitcher plant—was introduced to British gardeners by Thomas Lobb, a plant hunter sent to Southeast Asia in 1843 by the ambitious Veitch & Sons nursery. Lobb explored the dense rain forests of the region in search of orchids, tropical plants desired by the most discriminating of British gardeners. He was routinely swarmed by insects, had his blood sucked by leeches, and tromped through downpours, bone-chilling mists, and perpetual mud. He discovered the pitcher plant while on the Malay Peninsula. Lobb went on several plant-hunting journeys, and he did bring back valuable orchids to the nursery, including the blue orchid, the fox-brush orchid, and the popular moon orchid.

As the professional plant-hunting business grew more sophisticated and competitive, particularly concerning the wealth of plants being turned up by amateur hunters, botanists were dispatched on increasingly specific missions. This was especially true in the case of Ernest Wilson, who in 1899, at the defining age of twenty-three, was shipped off to China by Veitch & Sons in search of the deciduous dove tree. Once Wilson arrived in Hong Kong, a British medical officer handed him a map showing the location of a single dove tree. The map encompassed a region covering thousands of square miles. Nonetheless, Wilson eventually found the tree and collected its seeds. Freed from the fear of failure, he went on to identify a variety of plants on his first expedition, including the kiwifruit.

Wilson arrived back in England in 1902 and the following year was sent to Tibet, sailing up the Yangtze River in China, to find the yellow poppywort. He once again succeeded, and he found a number of other plants too, including the prized regal lily. Upon his return in 1905, Wilson delivered to his employers some 2,400 specimens and seeds of more than 500 species. He made subsequent trips to China and, even after breaking his leg in two places while trekking through the mountains, collected specimens and seeds of numerous plants. In addition to his other discoveries, he introduced to Western botany the paperbark maple, the rosy dipelta, and the candelabra primrose.

Similar to Fortune, George Forrest, who worked for the Royal Botanic Garden in Edinburgh, also went to China. By the time of his first trip in 1904, there was a wealth of flora flowing in from the Far East, and his initial benefactor, merchant Arthur Kilpin Bulley, wanted in on the excitement. Forrest voyaged to the Tibetan border on the northwestern edge of China's Yunnan Province, an unexplored wilderness for a plant hunter. Unknown to him, swelling beneath the rugged tranquility of the landscape was political upheaval. Tibetan priests, lamas, ruled the region by force and were fed up with the Chinese and the British, who were encroaching on their lands and culture. About the time that Forrest arrived near the Tibetan border, the lamas had taken it upon themselves to exterminate foreign intruders from their villages. When he learned that the assailants were descending upon the mission where he was staying, he and the sixty other people there, including women and children, evacuated immediately. The group was eventually attacked, however, and Forrest was the only one to have escaped with his life. He fell in with a group of Lisu peoples in a nearby village, and after an extended period of moving by night, under the cover of darkness, and managing only scant rest by day, starvation and exhaustion began to take their toll. At one point, he stepped on a bamboo spike that put an excruciating hole through his foot. Yet, despite all these struggles, Forrest managed to escape. In the process of the ordeal, however, he lost all of his possessions. While he lamented over the 2,000 specimens that had gone to waste, like any plant hunter worth his salt, he realized the need to move forward. He was determined, he was a naturalist, and he was a Scot.

Once safe from the terror of the lamas and healed physically, Forrest's passion for exploration and for learning about nature's wonders was reawakened. In 1906, after identifying new species of rhododendron, lilies, and other flowering plants and collecting a variety of seeds, Forrest returned to England. He visited China on six other occasions in the years that followed. Several of these later expeditions were supported by the Rhododendron Society, which was established in 1915. This connection with the society led to Forrest's introduction of a variety of rhododendrons to British gardening. He was also known for his introduction of several tea plants in the *Camellia* genus and of the brilliant blue Chinese gentian.

One of the last great plant hunters was Frank Kingdon-Ward, an Englishman from the county of Lancashire, perhaps better known for its cheese than for its botanists. Kingdon-Ward shared that most basic love for nature known to modern scientists like Wilson and to conservationists everywhere. At the age of nineteen, he traveled to Shanghai, where he became a teacher. His true interest, however, rested in exploration and the study of natural history. In 1909 he left his teaching post temporarily, in favor of joining a zoology expedition across western China. He was familiar with the expeditions of Hooker and enthusiastically took to helping American Malcolm P. Anderson collect animal specimens. Along the way he discovered several new species, including a couple of shrews and a mouse.

In 1911, back in Shanghai, Kingdon-Ward was contacted by Arthur Bulley, the merchant who had funded Forrest's initial voyage to China. Bulley enlisted the young man as a plant hunter, despite his inexperience. Kingdon-Ward took the job and shortly thereafter left for the Yunnan Province, heading to a village called Ta-li (Dali). Similar to Forrest, he fell victim to the political instability of the region, periodically finding himself fleeing from possible invasions and attacks. Still, he grew to love the Yunnan countryside, which was rich in wildflowers. And, like those who had come before him, he would make multiple trips to China in search of new and exotic plants to bring back to his homeland. Though he is best known for his introduction of the Tibetan blue poppy, Kingdon-Ward also delivered to Britain several species of rhododendron, *Primula*, and the popular wintergreen cotoneaster.

After Kingdon-Ward, support for plant hunting in distant lands waned, as many came to believe that all the best and most important plants had been discovered. By the time Kingdon-Ward was on his earliest journeys through China, it was already clear that highly sought after species were in peril. Between the mid-1800s and the early 1900s, dozens of plants succumbed to extinction, including species in Africa, the Americas, and Asia. Many of these losses were the result of agricultural expansion and deforestation, which reduced the native habitat necessary to support their survival, and natural factors, such as changes in local climate and fires.

The decline in interest in plant hunting occurred around the same time that interest in the development of plant-based pharmaceuticals dropped off. Suffice to say, in the early twentieth century, the world was changing, and rapidly. New technologies were emerging—from automobiles to air conditioning to instant coffee. Advances in chemistry were in the process of rendering plants unnecessary to the manufacture of useful drugs. Only in recent years have scientists realized what we stand to lose in plant diversity by clear-cutting forests and destroying animal habitat. Animals require specific habitat, which can be supplied only by plants and the pollinators that support them.

The discovery of exotic plants and the domestication of wheat, barley, and other crops were certainly important advances in human culture and civilization. But the human interaction with plants, especially their use medicinally, began long before gardening and plant domestication. The human impulse to explore and connect with nature is arguably among the oldest of human behaviors, underlying not only the ancient discovery of medicinal plants but also the far earlier discovery of plants as sources of food by other hominid species, a discovery that is predated even further by the use of plants as food by our ancestor primates.

But to what degree are human interactions with plants driven by instinct? And how far back in time can we travel before we run out of evidence in our quest to better understand how our ancient ancestors used plants? Studies of animal behavior in the wild have been eye-opening in the search for explanations of plant usage by prehistoric humans, and other research too has indicated that the use of plants as sources of food by our species presumably was not a sudden revolution. Only in rare instances, such as when people were suffering from hunger or poor nutrition due to the rapid loss of their primary food sources, or perhaps even out of random curiosity, were they likely to have serendipitously discovered the food value of certain plant species. In these cases, it was probably instinct and necessity that drove them to engage in trial-and-error plant ingestion. In all other cases, the ways in which humans used plants, as well as which plants people were drawn to, were the products of a very gradual process of primate evolution.

Observation of the natural world provided early humans with a basic knowledge of life, including patterns and cycles, which could then be incorporated into their daily lives and societies. These patterns and cycles were known to other life-forms long before they were ever known to the first humans.

Unfortunately, scientists' knowledge of the interactions between prehistoric humans and plants is restricted by limited evidence. The very term *prehistoric* precludes the existence of any written records, so scientists must rely on various combinations of details to draw conclusions about these early interactions. The existence of carbonized plant remains is particularly valuable for dating. Other sources of information include evidence turned up from archaeological studies, an understanding of the oral and medicinal traditions of native peoples, and knowledge about current plant distribution and use.

Information about the ways in which prehistoric humans used plants can be gleaned using other approaches as well. Studies of the dietary habits of prehistoric humans often involve the analysis of coprolites, or fossilized excrement, found in caves and geographical areas once inhabited by humans. Modern technology has provided scientists with the ability to analyze the DNA content of coprolites, picking out genetic sequences that are unique to plant chloroplasts and matching these sequences to known sequences stored in databases. In the case of coprolites found in Hinds Cave in southwestern Texas, this molecular approach revealed that the human inhabitants of the cave ingested a variety of plants, including buckthorn and a type of desert plant (a species of *Fouquieria*, possibly ocotillo). Based on existing knowledge of buckthorn and ocotillo, these plants were likely used for medicinal purposes rather than as sources of food.

The human inhabitants of the cave, who may have lived there as many as 11,450 years ago, were probably self-medicating with plants.

Even within the time frame of recorded history, the discovery of the medicinal properties of certain plants still in some cases occurred accidentally. Plants realized to possess qualities that favored their use to treat human ailments were likely then incorporated into traditional practices. To find plants, people had to wander, to explore the environment surrounding them. Among the places in the world with the most prolific recorded histories of ancient plant use are India and China.

India is a land containing a wealth of plant diversity, as European plant hunters discovered in the nineteenth and twentieth centuries, and it certainly presented an excellent opportunity for the discovery of medicinal species by ancient peoples. Some 8,000 different types of plants found in the region—from the neem tree (*Azadirachta indica*) to the camphor tree (*Cinnamomum camphora*) to *Aloe vera*—have been said to possess healing qualities. A significant number of these are used routinely by Indian physicians, and several that are used as traditional medicines have been studied by modern scientists in efforts to identify new bioactive compounds to serve as leads for drug development.

Many of the medicinal herbs used by ancient healers also were used as spices, added to various traditional food dishes. In the Cardamom Hills of India, tropical evergreens serve as vital shade trees for the area's namesake spice, cardamom. Other spices native to the hills include nutmeg and lemongrass, while a variety of plants, including tea, bamboo, and coffee, are cultivated there. The seedpods of cardamom have a rough texture and a strong aroma. In traditional Ayurvedic medicine, cardamom was used to treat upset stomach, hemorrhoids, and bronchitis.

For centuries cardamom was harvested from the plants native to India. However, as spices were traded and spread throughout the world, being used increasingly in the preparation of foods outside of India, the locals took to cultivation, growing the plants specifically for the harvesting of cardamom pods. This took place sometime in the early nineteenth century and led to cardamom's becoming one of the world's most expensive spices. Today, the plant is cultivated in its native habitat using a system of land management known as agroforestry, which involves growing the spice beneath and around trees on plantations. The shade provided by tree cover is welcomed by cardamom. However, it seems that not all shade trees are created equally. Furthermore, deforestation and the loss of plants have made cardamom cultivation difficult. Shade gaps between trees and less-than-optimal light filtering by certain tree species have hindered cardamom regeneration from one season to the next, causing lost crop yields in plantation areas. This in turn has led to the replacement of cardamom with crops such as tea and coffee, which are easier to grow and are more reliable from season to season. To grow these replacement plants, however, land must be cleared.

At the same time that Ayurvedic medicine was flourishing in India, traditional medicine was becoming firmly established in China. Traditional medicine in China, like that of India, was influenced not only by spiritual connections with the natural world but also by physical geography—the shape and character of the land and environment. Modern China is a land that has been impacted by the overgrazing of livestock, mining, deforestation, overpopulation, and the construction of dams. These factors have affected the natural ecosystems of the country. In one region in particular, the Loess Plateau, a mere 8 percent of the land remains untouched by humans, and it likely will not remain so for long, despite pressing conservation efforts. A contributing factor to the downfall of biodiversity in China has been the overharvesting of medicinal plants. The country's biodiversity has been reduced further by the hunting to near extinction of animals whose parts are highly valued in traditional medicine.

China was not always in the perilous state of loss in which it finds itself today. The history of its medicinal system certainly points to a land far richer in spirit and feeling than the industrialized attitude it has assumed. Much of Chinese medicine relies on an herbal tradition, and as such, it consists of many thousands of plant remedies. For millennia the Chinese have collected and cataloged all the plants with medicinal properties that they knew of. The people of ancient China were scientists in the same way that the people of ancient India were scientists. They experimented and used trial-and-error methods to determine which plants had healing properties and for which diseases each plant was best suited as a form of treatment. The ties between the Chinese and the natural and philosophical realms were very strong. Modern China seems much more analytical, and the drive for technological advancement is very strong. In most instances, this technological progression has been stronger than human biophilia. But interestingly, although the innate love of nature in this distant land, like many other places in the world, appears to have weakened over the course of generations, the knowledge of traditional medicines has survived with tremendous resilience, especially considering the significant movement toward conventional medicine that was encouraged by the government in the early twentieth century.

The plant remedies of India and China have and continue to be of great value to many people. Plant-based remedies constitute 85 percent of the medicines used in traditional practice, and some 80 percent of people in Asia and Africa depend exclusively on traditional forms of treatment. However, medicine has advanced by leaps and bounds in economically advanced countries. These places now serve as vital sources of medicines and technologies that are needed in impoverished lands, where most of the world's people afflicted by disease live.

As ethnobotanists and ethnopharmacologists dig deeper into the histories of the people and the plants behind the ancient systems of healing in India and China, as well as in other ancient societies such as those of Central and South America, we can anticipate a greater merging of the interests of conventional and traditional medicine. The harmonization of these two approaches stands to significantly advance the practice of medicine on a global scale. But for this to happen, it requires general acknowledgment that plant medicines must be subjected to standardization not only to meet the demands of modern drug development, but also to facilitate the discovery of compounds that could lead to medicines for currently untreatable diseases and to enable these medicines to benefit people worldwide. Those who practice and support conventional medicine must recognize and respect the art of ancient healing as a form of medicine that, globally, billions of people depend upon. This is no small feat, and by far the largest obstacle in our way is the disharmonic relationship between humans and nature in the modern era.

The studies of plant use in prehistoric and ancient human civilizations and the history of plant-hunting expeditions provide real evidence of humanity's relationship with nature. Our ancestors endeavored to connect with nature, whether for means of survival or for uncovering the beauty inherent in the natural world. Today we continue to seek contact with nature, and we can find beauty in everything from flowers to birds to bugs. But our modern perception of beauty arguably stems more from the human drive to develop a systematic knowledge of things than from the human love of nature. The former is an aspect of human behavior that is driven by technology, which fascinates us to no end, whereas the latter is driven by the threads of a severely attenuated biologic program. The human technological drive is at the root of all the things that we invent and discover, and it characterizes a side of human intelligence that represents our desire to build and our conquest of nature.

Yet, our survival depends on both nature and technology, and so we need to coexist with nature. This means living in fragile balance with all other forms of life on Earth, an element of humanness that reminds us that we are part of the animal kingdom. We have the ability to rediscover the role that nature fulfills in our lives, and we can strengthen the human bond with nature and reinforce an effective conservation ethic. But this relies on the hypothesis that our biophilic genes can be stimulated and expressed to a greater degree by our interactions with nature. The more frequently we interact with the natural world, the more sensitive these genes are believed to become, such that we will more readily identify natural dangers and will be more aware of the elegance and beauty that pervade the world around us. According to the biophilia hypothesis, these genes will then be passed on in this "primed" state to subsequent generations, providing our offspring with a useful degree of sensitivity to natural stimuli. If this really is the case, for biophilia to be truly reawakened in our species, we must connect with animals and plants and with the outside world in general, which means that we must disconnect from our couches, our computers, and our televisions.

4

In Earth's Garden

Madagascar periwinkle (*Catharanthus roseus*).

ONE OF THE MOST AMAZING things about nature is the vast body of information that it contains. Humans have been studying and exploring the natural world since the dawn of our species' evolution, and only in the last several centuries have we achieved a basic understanding of nature's most fundamental processes, such as photosynthesis and natural selection. Knowledge of who we are as humans and where we fit within the evolutionary tree of life is now expanding rapidly. Our most practical knowledge of nature, however, has been imparted by our individual interactions with it. And ever since the time when humans first encountered nature, this body of knowledge has grown continually. It has extended into our daily lives and shaped human cultures.

Gardens are nature's infiltration into society incarnate. They at once embody both the human dependency on and our fascination with plants. They are our attempt to bring nature to us, for sustenance and tranquility. Nurturing flowers, vegetables, trees, and herbs is an endeavor that transcends ethnicity, language, and geography, that offers a glimpse of what it means to be human. And further, large or small, public or private, all our gardens are united in the sense that they are made up of Earth's vegetation. It is this vegetation that forms the basis of the largest garden of all—Earth's garden. Jungles, savannas, wetlands, deserts, farms, yards, and parks fall within its boundaries. We are but one of millions of species that live in this garden, and our activities contribute to its growth and beauty as well as to its deterioration and destruction.

Our individual gardens are outlets to nature, where we can fulfill basic human needs, such as spending time outside. A garden may consist of a bed of wildflowers, a container of herbs, or acres of planned parterres, dotted with fountains, aesthetically trimmed hedges, and other features. Many gardens that are open to the public are exquisite in their design and landscaping and contain a wide variety of plants, many of which have been specially bred and some of which are very old. But thanks to the invention of container gardening, land is not necessarily a prerequisite for surrounding oneself with vegetables and flowers. This is just one of the factors that lends a sense of universality to gardening—anyone can do it.

But, as simple as it can be, gardening can also quickly swell into an extraordinarily complex and cerebral hobby, involving knowledge of species names, genetics, and soil chemistry. The vast quantity of literature on gardening collectively tells a story about how and why we use plants. The cumulative history of human experience with growing plants underlies modern cultivation and explains the multitude of other ways in which we use and interact with plants. From this history, there emerges a coherent picture of how our species has grown from filling a basic role in nurturing Earth's garden to serving as its dominant species. We have evolved to the point where, in the modern era, we possess the ability and the technology to both nurture and permanently harm its vegetation. We are able to distinguish between these two actions, and we know how the former is accomplished and how the latter can be prevented.

Long before there were packaged rooting hormones, water wands, and grow lights, there existed only a primitive approach to gardening. It was, and to some extent still is, a trial-and-error process. It is also highly variable geographically and culturally, particularly concerning motivation and design. Planned gardens were probably in existence by at least as early as 2500 BCE. In the Euphrates valley of the Mediterranean, proof of early organized gardening efforts is supplied in written records and paintings, wood carvings, and inscriptions.

The most convincing early evidence of gardening in Egypt dates to around the time when Snefru reigned as the first king of the Fourth Dynasty (c. 2575 to 2467 BCE). On the walls of a tomb dating to that era, archaeologists found inscriptions describing a garden and vineyard, complete with cultivated trees and a lake. Popular plants at the time included sycamore fig trees, tamarisk, and persea. The Mediterranean fan palm, the leaves of papyrus, and the flowers of the blue Egyptian water lily were used to adorn pottery, buildings, and public spaces. In the centuries that followed, Egyptian gardening adopted an increasingly geometric attitude, with organized tree-lined avenues, planned gardens, and rectangular pools.

Even in ancient Egypt, humans embellished their gardens with plants that were not native to the soils in which they were sown. The incense trees that Queen Hatshepsut ordered botanists to track down sometime around 1495 BCE were in fact intended for decorative use around the western Thebes temple, Deir el-Bahri, and for burning as aromatics. Predating Hatshepsut's incense trees appears to have been the import into Egypt of mandrake and pomegranate. Later, Asian cornflower, red poppy, and opium poppy were introduced. The various kings and queens appear to have indulged in the importation of plants and in the planning of elaborate gardens.

The temple in the ruins of Tell el-Amarna, where King Akhenaton and his wife Nefertiti resided in the fourteenth century BCE, is said to have possessed a garden designed in honor of Aton, the sun god. Archaeological studies have indicated that date palms, doum palms, and pomegranate were planted there.

Features to support plant cultivation, such as canals and irrigation systems, were also developed in ancient civilizations. Canals to shunt water from the mighty Euphrates and Tigris Rivers to the parched desert plains of the Land Between Rivers, Mesopotamia, enabled agriculture to take root, expanding food production to support the region's growing settlements. These human-made water passageways were in place by about 3000 BCE, having been introduced by the Sumerians, who also created the first system of writing. The larger region of Mesopotamia itself was also known for the *Epic of Gilgamesh*, an odyssey about the king of Uruk, a city described as being one-third garden and home to one of the Seven Wonders of the World—the Hanging Gardens of Babylon.

In 689 BCE, Babylon, which was then the capital of southern Mesopotamia, was destroyed by the Assyrians, who quickly established their own empire in the region, situated around the city of Nineveh. The Babylonians later united with both the Scythians, a nomadic Iranian people, and the Medes, an Iranian people who occupied Media, a land on the bank of the Tigris opposite Mesopotamia. This unified front captured Nineveh and reclaimed control of the empire in 612 BCE. The new Babylonian king, Nebuchadnezzar II, rebuilt the capital of Babylon, and it is there that he is said to have designed the Hanging Gardens as a gift to his wife, the Median princess Amytis. The writings of Greek historians provide descriptions of the Hanging Gardens, and according to their accounts, they considered the gardens to be one of the Seven Wonders of the Ancient World. The site still holds this title today, though physical evidence of the gardens is scant.

The princess's Persian roots probably inspired the rooftop layout and types of plants grown in the Hanging Gardens. The rooftop design, with long vines trailing down from terraces and balconies, made the gardens appear as though suspended in air, producing the "hanging" effect. Greek historian Diodorus Siculus has supplied us with what is considered to be, more or less, an accurate portrayal

Mesopotamia and the locations of Babylon and Nineveh, c. seventh century BCE.

of the gardens' design. He described a sloping landscape and the construction of level gardens, arranged in tiers as the hill climbed upward. Each tier was densely covered with trees and other plants. There was also supposedly a machine capable of moving large quantities of water from the river to the gardens. Diodorus refers to this as a sort of hydraulic system, used to draw water up from the Euphrates to each of the terraces lying above the river. This system was probably similar to the Archimedes screw, a device believed to have been designed by the great Greek mathematician of the same name for the purpose of conveying water from the holds of ships. The screw also found use as an irrigation device, moving water from low areas to high areas. The Hanging Gardens of Babylon are believed to have been among the first gardens to use such an elaborate irrigation system.

While there is evidence supporting the existence of the city of Babylon, in the form of cuneiform inscriptions, foundational structures, and other archaeological data, whether or not the Hanging

Gardens were real is another matter. That the ancient Greeks wrote about the terraced gardens of the capital seems to prove their existence at one time. The information penned by Greek historians is derived in large part from the writings of Berosus, a priest in Babylon who lived sometime around 290 BCE. Passages that he composed, such as his description of Nebuchadnezzar having built the gardens for his wife, were quoted by others after him. Other, generally authoritative accounts include those of Greek historian and geographer Strabo, who described the size of the walls surrounding Babylon as being thirty-two feet thick. He also described the four-sided shape, arched vaults, and baked brick terraces. Greek physician and historian Ctesias wrote a presumably reliable account of the gardens in his work *Persica*, a twenty-three-volume history of Babylonia and Assyria. There are no solid remnants of the gardens, but the writings have sufficiently captured the human imagination. Many modern gardeners aspire to the creation of such a captivating expanse.

The idea of landscaping with conspicuous and fragrant flowers and luxurious fruit-bearing trees naturally passed from the Babylonians to the ancient Greeks who came to occupy their lands. The Hellenistic period of Greek history began when Alexander the Great took control of Babylon in 331 BCE. His empire ultimately included Africa, the greater part of Europe, the Middle East, the Mediterranean, and Persia. Hellenistic gardens incorporated different elements of these cultures and included the traditions of Byzantine gardening. Byzantium, a Greek civilization, was established sometime in the eighth century BCE along the Strait of Bosporus.

Similar to the way in which the *Epic of Gilgamesh* provided scholars with a glimpse of the Babylonian relationship with gardens, ancient Greek literature gives us an idea of the great variety of plants cultivated by people in early Mediterranean civilizations. In the *Odyssey*, Homer describes the garden of Alcinous. Believed to have been located on the island of Corfu in the Ionian Sea, the garden was embellished with pomegranate and apple, pear, fig, and olive trees. The variety of vegetation described and inclusion of fountains and a hedge border are believed to describe Homer's ideal landscape, though his elaborations may also allude to the typical farmhouse gardens already in existence during, or even prior to, the period when the work was written.

Plants in ancient Greece served multiple purposes, including as sources of medicines, shade, inspiration for paintings and decorations on pottery, and ornamentation in small domestic gardens and areas of public gatherings. The famous Academy established by Plato outside the walls of Athens had gardens and groves of olive and plane trees, which offered places for contemplation. Greek philosopher Epicurus also took to gardens, building his only a short distance from the Academy. Epicurus's garden, closed to the public, came to represent a place from which emerged hedonistic philosophy. Within Athens, trees such as poplar, cypress, holm oak, stone pine, Kermes oak, and Grecian fir were planted along avenues and in groves in the city's market area. The Temple Garden in Athens is one of the first examples of an organized attempt at garden design in ancient Greece.

The physical labor of gardening was considered by the ancient Greeks to be the work of slaves. Persons of high rank often simply indulged in the extravagance of gardens, expecting the work to be carried out by their laborers. There were exceptions, however, including the Persian prince Cyrus the Younger. As time went by, plant cultivation became viewed as a skill. Students at the Lyceum, a school and public assembly hall established by Aristotle in 335–334 BCE, were taught seed germination and grew plants for the great Alexander.

Many similarities have been observed between the luxurious gardens of the Hellenistic period and the pampered gardens that followed in ancient Rome. Gardens served many purposes in the city and surrounding areas, and the writings of Cato, Varro, Palladius, and Pliny the Younger reveal that the Romans possessed a deep appreciation for natural landscapes and a desire to live submerged in an atmosphere of extravagant and exotic plants. Villas were often designed in conjunction with plans for a garden, and peristyle gardens, in which plants adorned the spaces within a courtyard, which itself was typically surrounded by columns, were common in ancient Greece and Rome.

Archaeological excavations at Pompeii and Herculaneum led to the discovery of various elements of Roman garden design, including the layout of flower beds and fountains. The gardens of Pompeii were well preserved under volcanic ash from the Vesuvius eruption of 79 CE. Other gardens that have been documented, either in Roman literature or in evidence from ruins, include the Gardens of Lucullus and the gardens at Hadrian's villa at Tivoli. While it is unclear which plants Hadrian had growing in his garden, it appears that each building in the compound had its own arrangement of flowers and trees, and several buildings had canals and fountains.

The Romans made use of a wide spectrum of plants and gave careful attention to cultivation. Manuals were composed specifically on farming and gardening and included a host of details on cultivation techniques, ranging from the best time of year to grow certain types of plants, to how to prune vines and trim hedges, to how best to maintain nursery trees. Brambles, roses, bay laurel, acanthus, magnolias, and evergreen shrubs are examples of plants that were incorporated into the layout of Roman gardens. And much the same way that other civilizations had imported exotic species, the Romans, too, introduced foreign plants to their own gardens, including the plane tree, which had been brought from Greece's Ionian Islands to Italy.

The concept of cultivating plants in an organized fashion was later adopted by the Islamic world. The gardens of Islam made exceptional use of existing knowledge of plants, which came primarily from scientific information recorded by the ancient Greeks that was translated into Arabic. Islamic gardens were heavily influenced by the sacred writings of the Quran. The purifying nature of water was emphasized in pools and irrigation systems, which often served symbolic purposes. The extensive trade networks that fed into the Middle East, from India, the Far East, and other regions, led to the introduction of a great variety of plants, everything from banana trees to roses, coriander, plane trees, and opium poppy.

The beauty and luxury of the ancient Greek and Roman gardens disintegrated following the collapse of the Roman Empire, which marked the beginning of the Middle Ages. As darkness settled over European society, interest in learning and science faded. Academic advance halted and, at some points, even regressed. The art of gardening very nearly disappeared. Ruined by several centuries of disorder, and with the pursuit of knowledge in the arts and sciences attenuated, many people turned to belief in the supernatural, forgoing opportunities to question the phenomena of the natural world.

During the Middle Ages, some of the most significant advances in the scientific knowledge of botany took place in monasteries, where there often existed a recognition that learning fulfilled an important place in human life. Although gardens were first used almost exclusively for food production, as centuries passed, a variety of medicinal and inedible plants were grown. The latter enhanced the beauty of abbeys. The scent of flowers and vegetation, the rustle of the breeze through leaves, and the buzz of insects flitting from plant to plant likely served as subtle reminders of God's omnipotence and desire to impart beauty on even the lowly world of earthly beings.

A great deal of medieval gardening was devoted to the growth of medicinal plants. In the seventh century, St. Fiacre, the patron saint of gardening, established a garden near Meaux, France, that became well known for its medicinal plants. Most medieval gardens, however, served a dual purpose—as a source of food and as a source of medicine. Plants more generally were viewed as resources, most evident in the clearing of forests for their wood, which was used to build just about everything except stone castles. In many ways, the treatment of plants and animals in the current era echoes this antiquated perception of nature existing in endless fashion for our use.

Many of the trees and flowers that were known to the ancients were also known to later medieval scholars. Almonds, figs, peaches, and pears had been introduced to northern regions of Europe by the Romans. Likewise, roses, lilies, and hemp were brought to Europe by earlier peoples. But while some medieval monastic gardeners possessed knowledge of plant grafting, pruning, and germination, most other people knew very little about plant cultivation, save for their experiences with species they may have tended in their home gardens. Plants introduced north of the Alps, including various types of citrus trees, which demanded greater attention and care, appear to have died out from neglect.

According to historical records, monastery gardeners actively engaged in seed trading and grafting, which helped enrich the vegetables and flowers they grew and suggests that monastery inhabitants knew a little something about plant cultivation. As the Spanish accumulated a knowledge of Greek and Middle Eastern history, they must have learned about plants and agriculture. The Arabs and Spanish established trade routes, and this is probably how plants such as irises and rosemary entered Europe. Slowly, gardens magnified from the simplicity of monastery kitchen gardens to more sophisticated pleasure gardens like the haven at the Nuremberg castle of Holy Roman Emperor Frederick II. Still, these gardens were nothing compared with those of the Romans and Arabs.

As trade with the Arabs expanded to other parts of Europe, the plants known to the Arabs acquired a greater presence in European gardens. Many species weren't suited for the cool northern climate. Date palms and olive trees struggled to grow. In the late 1100s Alexander Neckam published *De Naturis Rerum*, which described herbs, flowers, and trees that could be grown in Europe. The following century, Albertus Magnus wrote about ways to arrange herbs and flowers within a garden. Magnus was an herb aficionado. Neckam seemed to gravitate toward ornamental plants, including violets, roses, marigolds, mandrake, and poppies. Between the two of them, Europeans were exposed to a new and fairly comprehensive literature on garden plants. The concept of the pleasure garden seemed a natural follow-on. Similar to Islamic gardens, "pleasure" gardens in the later medieval era were both places of religious symbolism, with different types of flowers serving as interpretations of elements of Christianity, and places of romance. Medieval gardens were gardens of Eden and gardens of Chaucer. They were solemn, and they were magical.

Gardens also became commercial entities. Fruit and vegetable trade existed in earlier civilizations, but the commercial gardens of the Middle Ages were like rudimentary versions of the large farms that now lie at the edges of cities and support food production for urban populations. In medieval Europe, expert gardeners oversaw large nurseries or gardens on the outskirts of cities such as London. Vegetables, flowers, and other plants were grown in large quantities and sold for profit at public markets within the cities.

By the time the Middle Ages came to an end, the idea of a small, protected garden adjoined to a castle or villa had grown in popularity. The rebirth of classical design inspired creativity, and large, lavish gardens were arranged to adorn the properties of the well-off. Arts and culture, so long suppressed by the stringency of the Middle Ages, flourished during the Renaissance, and gardens, places for plants, also became places of art. In Italy, drawing on classical themes, garden architects reintroduced the pergola, on which climbing plants were grown and formed a roofed passageway. Statues, fountains, pavilions, and grottoes added entirely new elements to gardens, and the practical garden of the Middle Ages became comparatively dull.

The profuseness of expense and the quest for a pleasure garden unlike any other drove Renaissance gardens to unprecedented extremes. The gardens of Belvedere Court at the Vatican, Villa Medici di Castello near Florence, and Villa d'Este in Tivoli embodied the audaciousness of Renaissance gardens. Villa d'Este, designated a UNESCO World Heritage site in 2001, was designed by architect Pirro Ligorio around 1550 and is the epitome of Renaissance culture. Its magnificent use of water, its statues, and its layout are like none other. As far as plants are concerned, however, it is actually rather unremarkable.

The Italian style spread across the Alps and into France, where many gardens were designed to enhance the flat landscapes surrounding them. The best examples of Renaissance gardens in France include those of Fontainbleau, Chateau d'Amboise, and Tuileries Palace. Across the English Channel, the Renaissance flair for gardens initially clashed with the traditionally Gothic architecture. Later, however, Italian garden designs were synthesized with Tudor and Elizabethan designs, giving rise to extensively planned gardens, such as those of Hampton Court in Herefordshire and of Nonsuch Palace in Surrey. These gardens retained an English romanticism, adorned with sculptures of birds and other animals, rather than the classical human statues found in Italy. By the Baroque period, the late 1500s to 1700s, however, garden design across most of Europe and England had become less about nature and more about the human manipulation of nature and theatrics. Gardens were built to impress and were increasingly less practical. In later centuries, many of these lavish gardens were perceived as stilted, their deficiencies in substance often thinly veiled by flashy waterworks.

In the later Romantic period of the 1700s and 1800s, there was a push for a return to natural gardens. This theme—the garden as it exists in a natural landscape, with little human persuasion—has been revisited at various times ever since. At the same time, the Enlightenment was upon Europe and Britain, and there ensued a tremendous acquisition of scientific knowledge of plants. Preceding this period, scientific interest in plants, particularly in the realm of medicine, had already made substantial progress, fueled by the work of the botanists in the New World and by the three German botany pioneers,

Otto Brunfels, Hieronymus Bock, and Leonhard Fuchs. Interest in plants themselves grew in parallel with the extravagance of Renaissance gardens. This was particularly noticeable in the increasing fascination with exotic plants, which were very rare in European and British gardens. Plant cultivation was in many ways only emerging from its infancy at the time, with complexities about soil chemistry and hybridization relatively unknown. Linnaeus's classification system did not appear until the mid-1700s, and species names of plants, if they were used, were likely to be different in different places of the world. But exotic plants did not need Latin names for people to recognize that they were unique. Their peculiar characteristics were prized. In the Renaissance era, they attained the then prestigious status of collectibles, in part forming the basis for the later interest in sending plant hunters on expeditions to track down new species.

Much of the botanical knowledge gained during the Enlightenment stemmed from earlier advances in the organization of universities in Europe. As the Middle Ages gave way to the Renaissance, scholars encountered the works of the ancient Greek naturalists, as well as the works of then contemporary botanists. Brunfels's *Herbarium vivae eicones*, Bock's *Kreuterbock*, and Fuchs's *De Historia Stirpium Commentarii Insignes* were studied, and the body of botanical knowledge expanded rapidly.

Freethinking in the Renaissance led to the general acceptance of the notion of the natural garden. Such gardens were based on the irregular features of natural landscapes, with less geometrically restrictive layouts than those observed in earlier designs. These gardens were less intense in terms of their demands on irrigation and manual labor. The concept of the "natural garden" now has various meanings, from fitting in with the natural landscape to involving the use of only native plants within their native habitats. But the fundamental idea is the same, that nature can help out. Natural gardening with native species makes use of adapted plant features, including thorns to defend against herbivores and specialized leaf structures to prevent the loss of moisture in plants native to dry climates. The cultivation process can be greatly simplified through reduced human intervention and emphasis on the use of native plants, which are naturally suited to the soil conditions, climate, and wildlife.

Gardening provides an excellent opportunity to connect with nature. Here, at a school garden in Brooklyn, New York, a US Navy volunteer and a student plant flowers. (Photo credit: US Navy photo by Photographer's Mate 3rd Class Gina M. De Jesus)

An important element of a natural garden is its potential to attract wild creatures. Native plants attract native animals, and more importantly, they can support these creatures. An exotic plant adapted for pollination by a single species of insect but displaced to a foreign habitat will not survive without human intervention. Exotic plants that do thrive and replace native flora force native fauna to seek food and shelter elsewhere. Large swaths of land, once rich with a wealth of plant life, but now covered with a neatly trimmed lawn and lined with foreign shrubs and flower beds of nonnative species, are devoid of native wildlife, generally with the exception of generalist species. So, while we may revel in the aesthetics and symbolism of lavish gardens, such extravagances often come at the expense of native animals and plants.

Most people who do their own gardening simply enjoy the experience of being outdoors. They like to watch their plants grow week by week and listen to birds calling and insects buzzing. These are the real attractions, and there is a certain appeal in growing native plants that support local animal populations. Gardening is a hobby in which the greatest pleasure is derived from a combination of being surrounded by the richness of plant life and of performing seemingly mundane tasks, like digging holes and pulling up weeds. There could be a real physiological connection to this too. Coming into contact with certain soil bacteria has been found to stimulate the release of serotonin in the brain, a molecule that is often the target of antidepressant drugs.

Our little gardens remind us of nature's wonders as they exist right in our backyards. National parks and other protected areas remind us of the beauty and importance of the larger sense of Earth's garden. Where the presence of humans is limited and nature is undisturbed, the interactions between plants and animals play out as their evolutionary histories intended, often leaving even the most casual observer intrigued.

In the current era of unprecedented expansion of human populations, the concept of the national park is one of the most important things that could have happened for the environment. Humans have long recognized the innate beauty of nature, but finding a way to preserve it for the long term was a separate matter. National parks were born in the United States in the late nineteenth century, at a time that paralleled the onset of the country's Industrial Revolution. Then, as now, people were calculating and thinking ahead. Where some saw progress, others saw a country with a growing population and an abundance of consumers ready to capitalize on the seemingly never-ending blanket of US territory.

The idea to set aside land within park boundaries originated with the problems of Yosemite. A short distance east of San Francisco, Yosemite is a land filled with mountains, waterfalls, glaciers, meadows, and a wide variety of plants and animals. It is a rugged place and traditionally home to tribes of Native Americans. With the arrival of white European settlers to the region between the 1830s and 1850s, however, the welfare of the landscape and its native inhabitants fell under threat of exploitation and destruction. One of Yosemite's first defenders was architect Frederick Law Olmsted, who oversaw the development of Central Park in New York City. Olmsted visited Yosemite in the early 1860s and was instantly enamored with its beauty. He decided to work with local conservationists to push for the land's protection under a trust. In 1864 Abraham Lincoln signed the Yosemite Land Grant, which put the region under state and public control. Olmsted was subsequently appointed to serve on a board of commissioners overseeing the lands of the newly declared public trust. He surveyed the region's boundaries, using his own money to support his work. He understood the need to preserve the lands, and beyond this, he envisioned a "wild park," one that could be enjoyed by future generations.

Olmsted's sentiments about Yosemite, which were shared by many influential artists and naturalists who had been to the region, had an important role in the development of not only Yosemite National Park but also the broader idea of a national park system in the United States. In 1890 Yosemite finally earned park status, becoming the second to be established in the country (Yellowstone National Park, established in 1872, was the first). In the time between its formation as a public trust and its designation as a national park, Yosemite experienced a significant amount of growth and attention. John Muir visited the region in 1868 and was an important supporter in its later elevation to park status. Painter Albert Bierstadt and photographer Carleton E. Watkins also made journeys to Yosemite. Bierstadt, inspired by the scenes around him, created "The Domes of Yosemite," "Valley of the Yosemite," and other works, which effectively captured the beauty of the land. Watkins's photographs were equally enthralling, and wealthy people from the East, enticed by these works of art, were suddenly curious about this land. They wanted to see it for themselves. And thus, even under the watchful protection of people like Olmsted, Yosemite was exploited. New hotels were built, stores were opened, trails were carved, space for public campgrounds was cleared, and tracks for the Central Pacific Railroad, the tourist artery of Yosemite, were laid. It was a large amount of activity in a short span of time, with the result that much of Yosemite has been altered by humankind. Even today park officials struggle with the sheer volume of people who visit each year. While economically the number of park visitors is beneficial, natural habitats within the park's boundaries suffer.

The first expanse of natural environment placed under US government protection was Yellowstone National Park, which was established on March 1, 1872. Yellowstone, tucked into the northwestern corner of Wyoming and edging over the north and west borders of the state into Montana and Idaho, was a vital first step in nature conservation. Whereas Yosemite was very nearly trampled by settlers and visitors, Yellowstone remained virtually pristine. Its lands had been known for generations only to Native Americans, who criss-crossed its valleys and mountains in pursuit of bison and in search of edible plants.

The first people to encounter Yellowstone other than Indian tribes were white explorers and trappers. The extraordinary character of Yellowstone must have astounded these early visitors. Its canyons, meadows, hot springs, waterfalls, and petrified trees earned the respect of all who traveled there. Prospectors explored the land decades after trappers had been there, but the ruggedness of many areas prevented extensive exploration. It wasn't until 1869, when three brave adventurers, David E. Folsom, Charles W. Cook, and William Peterson, willingly ventured into the depths of Yellowstone, to places where people had traveled before but were rumored to have never returned, that more became known of the mysterious land. One year later, Folsom, Cook, and Peterson emerged from Yellowstone, loaded with details about its natural features. That same year, a surveyor of the Montana Territory launched a second expedition, which confirmed the discoveries of the three and resulted in the publication of a number of articles about Yellowstone.

The head of the US Geological and Geographical Survey of the Territories, Ferdinand Hayden, learned of these expeditions and in 1871 embarked on what became known as the Hayden Survey of Yellowstone. The survey was far more comprehensive than the ones before, and a team of naturalists, botanists, zoologists, and artists recorded a significant amount of new information. Hayden and his team realized that the innate value of Yellowstone could never be assigned any monetary or material meaning, not even in a country in the process of an industrial revolution. They appealed to Congress to have the land set aside, and in 1872 Ulysses S. Grant did just that, establishing Yellowstone National Park. In its early days Yellowstone served as a sort of testing ground, in which park officials were faced with the learning curve of overcoming the construction of roads and the establishment of visitor centers and park rules. Throughout the years, however, Yellowstone's wilderness, which encompasses nearly 3,470 square miles (2,220,800 acres), has remained almost entirely undeveloped.

A major focus of organizations such as the National Parks Conservation Association, which was created in 1919 by the same group that founded the US National Park Service, is finding ways to maintain the biodiversity of natural habitats and prevent the loss of species. But biodiversity is more than the number of different species found within an area. It is defined by the presence of diversity on a higher level of classification—the variety of genera, families, orders, all the way up to kingdoms—and by the presence of genetic diversity within populations of individual species. It is also variation between Earth's ecosystems, with many different systems providing homes to many distinct sets of living organisms. At the level of individual ecosystems, maintenance of the variety of life is largely dependent on size of geographical area. And by this measure, protected areas represent an effort to preserve what is left of an ecological community that was once much larger and much more diverse than it is now.

Within isolated geographical regions, which are separated from other areas by features such as mountains or bodies of water, there often exists a set of endemic species. These species can be found only within that isolated region; they are found no place else on Earth. Olympic National Park in Washington is a useful model for understanding endemism and the influence of geographical phenomena on biodiversity. It has more than 20 endemic plant and animal species, and it and the entire Olympic Peninsula on which it sits are home to as many as 1,450 species of vascular plants. The region's endemic species exist there and no place else because of the peninsula's mountains and relative continental isolation. Movement of Ice Age glaciers formed Puget Sound, which separates the eastern edge of the peninsula from the interior land, and the Strait of Juan de Fuca, which separates the northern edge of the peninsula from Vancouver Island. To the peninsula's west is the Pacific Ocean and to the south are Grays Harbor and North Bay, leaving only a small flank of land connecting it to the rest of the North American continent.

Most of the endemic plants found in Olympic are wildflowers, such as the Olympic mountain daisy, the Olympic violet, and the Olympic bellflower. The wide diversity of plants found within the park and on the peninsula is largely a function of elevation and climate variation.

The Hoh Rain Forest, on the west side of Olympic National Park in Washington State, is a rare example of a temperate rain forest. (Photo credit: Jeremy D. Rogers)

Elevations range from sea level to 7,979 feet at the peak of Mount Olympus, and annual precipitation ranges from 140 to 170 inches in the Hoh Rain Forest on the western side of Olympus to about 25 inches near Port Angeles, which lies along the Juan de Fuca Strait. Temperatures are generally mild, though they may dip below freezing on the mountaintops. The mild temperatures and precipitation support the growth of many different kinds of plants and allow trees to grow undisturbed by frost or drought for long periods of time. The park's green landscapes of trees, mosses, and ferns at lower elevations and its vivid purple, white, and yellow flowers that emerge at higher elevations in the summer are heavily influenced by local climate.

Olympic, just like all other protected areas, has a long natural history and a detailed political history of how it became a protected area. The trees of the peninsula are great sources of pulpwood for paper products, and Sitka spruce is highly sought after for certain types of musical instruments. In the late 1800s, giveaway land laws in the West let people snatch up public lands for very small amounts of money. Timber companies acquired vast expanses of forested areas. In the early 1890s, land giveaways were finally curtailed, and forest reserves were established. In 1897 Grover Cleveland, much to the dismay of the timber community, declared that some 2,188,800 acres were to be protected within the Olympic Forest Reserve.

Initially, the forest was closed to logging. But a vote passed the following year permitted state senators to lobby for the use of forest reserve land for agricultural purposes. The timber industry moved swiftly, and in Washington a new state senator was elected, one who supported logging and who reclaimed forest reserve lands for the timber industry in the name of agriculture. This meant not only that logging, under the guise of agriculture, would resume, but also that livestock would be allowed to graze the forest's meadows. In 1901, giving way to political pressure, the size of the forest was reduced, by presidential order and in the interest of timber production, to 721,920 acres. Several years later, in 1909, Theodore Roosevelt, realizing that the local elk populations needed additional protection, established the Mount Olympus National Monument. The Roosevelt elk was named for him, and because of his interest, herds of these animals still graze the forest today.

But the monument title was not sufficient to fend off the forces of logging, and once again the size of the protected area was reduced, this time to 600,000 acres. Decades of arguments ensued between politicians, conservationists, and preservationists. The latter, led by influential naturalists such as Muir, pushed hard for the formation of a national park to forever protect what was left of the natural landscape. Finally, in 1938 Franklin Roosevelt, who had made the long voyage west to see the land, designated it Olympic National Park. Even then, however, determined timber workers logged trees illegally within the park boundaries, and eventually, populations of Sitka and other trees had been logged so extensively that they were in danger of extirpation. Conservationists stepped in, and preservation gained local and national support.

Despite all that had happened around its borders, the park itself became a UNESCO Biosphere Reserve in 1976 and a UNESCO World Heritage site in 1981, in recognition of its rich diversity of ecosystems, life, and geology, and for its pure beauty. Around the same time, tree farms had been established outside park boundaries, which limited the expanse of land given over to logging. Today, a drive around the west side of the park reveals young stands of trees growing up between naked strips of ground where their ancestors have been cut down. It is an oddly realistic picture of how we are learning to coexist with nature, respecting its beauty and life and working to find ways to use its resources in a sustainable manner.

Species interactions are a pervasive and necessary feature of nature. Relationships between species exist for fundamental reasons, with the most obvious being survival or gaining a survival advantage. Many relationships are a result of subtle dependencies that connect plants to animals and topography. These dependencies also mean that many plants and animals have supporting roles in processes that are not directly related to their individual survival but are crucial to the maintenance of an ecosystem's function. Gaining a more extensive knowledge of these indirect relationships in nature has become extremely important in understanding how human activities impact the natural world. In many cases our actions indirectly trigger the decline of animal and plant populations. Such declines in turn affect human societies—for example, causing the loss of plants that could serve as sources of medicines or reducing genetic diversity among plants and animals to the point that pathogens more readily spread throughout ecosystems, further shrinking species populations in the process.

The idea of the "national park" is a phenomenon that reaches much further than the United States. International organizations such as the World Commission on Protected Areas (WCPA), part of the International Union for Conservation of Nature and Natural Resources (IUCN), help countries worldwide find ways to protect and preserve natural habitats. According to WCPA, about 5,115,850 square miles (3,274,144,000 acres) were harbored within the 30,000 protected areas in existence in the world by 2000. Many parks are constantly working to make ends meet, so there is a special significance in the ability of a park to successfully strike a balance between natural preservation and economic stimulation. If people did not visit protected areas, maintaining them would be difficult, if not impossible. But too many visitors can very quickly rob nature of its health. Crowds trampling off designated paths or people improperly disposing of trash or carelessly feeding the wildlife from their hands can cause real destruction to a park's natural ecosystems.

National parks visitors are there for many reasons—solitude, relaxation, a refreshing excursion into the wilderness. Preserved natural areas also provide us with a sense of our heritage and our natural history. Our sense of connectedness with the environment itself is difficult to describe and seems to share an intimate association with biophilia. But we feel something when we are in nature, and especially when we are in the physical domain of national parks, nature reserves, and protected wildernesses, where the human presence is small.

Our awareness of and interactions with nature enable us to cultivate meaningful relationships with the world around us. Because we possess knowledge of nature and the ability to protect and conserve it or to dominate and harm it, reestablishing this type of meaningful relationship is fundamental to the future well-being of Earth's biodiversity. The biological world is dauntingly complex. Yet, in understanding the many components fundamental to interactions between different life-forms, nature grows all the more amazing, simultaneously revealing her resilience and her fragility. In achieving an awareness of nature, standing on the edge of the Grand Canyon becomes less a tourist excursion and more a meaningful experience. That life on Earth is far more sensitive than we may tend to think has become increasingly evident from the way human societies interact with the environment.

Earth's biodiversity is vanishing. Understanding how and why is important for reasons much broader than simply the discovery and development of new drugs to treat our ailments. Vanishing plants, vanishing animals, and vanishing ecological communities render all life on our planet susceptible to disease and extinction. In the worst-case scenario of global struggle against disease, it would not matter how many or what kinds of drugs we have available. Treating a sick world takes more than drugs alone.

5

Vanishing Life

Gumwood (*Commidendrum*; top), St. Helena plover (*Charadrius sanctaehelenae*; left), and *Trochetiopsis* (right), endangered species of St. Helena Island.

THE HUMAN SPECIES is a recent development in Earth's history. Our lineage, the hominin lineage, comprising extinct human relatives and extant humans, arose between 5 million and 8 million years ago. Compared with creatures like the platypus and the spiny anteater, the last of the egg-laying mammals, whose lineages diverged between 20 million and 50 million years ago, and the first vascular plants, which appeared an estimated 410 million years ago, our species has existed but a moment in Earth's history. So although we may consider ourselves to be an extraordinarily intelligent species, we are among the planet's youngest, and consequently, our cumulative amount of experience with nature is relatively small. At times throughout our history this inexperience has revealed itself in uneasy fashion, in our mindlessly killing wild animals and tearing plants out of the ground, in our letting self-interests control the fate of our environment, in our throwing tantrums because we are not permitted to drive our off-road vehicles across tender wildernesses.

Many do not realize that we are in the midst of a mass extinction episode, the sixth such event in Earth's history. Mass extinction is defined by the loss of multiple groups of organisms, particularly genera and families, within a brief span of geologic time. Such events are a part of Earth's natural history, and much of what is understood about extinction in the modern era has come from studies of the fossil record. This meticulous, though incomplete, account of the history of life on Earth tells us what happened during the five earlier mass extinctions, which included the Ordovician extinction, the Late Devonian extinction, the End Permian extinction, the End Triassic extinction, and the ensuing End Cretaceous (or Cretaceous-Tertiary) extinction, which did away with many species about 65 million years ago. All the while, persisting between and among these five major events, there was a continuous level of "background" extinction, characterized by the periodic loss of comparatively few species here and there.

But whereas the five previous mass extinctions have been attributed to factors such as climate change, movement of Earth's tectonic plates, sea-level fluctuations, volcanic activity, and the occasional freak asteroid, the current, sixth mass extinction period appears to be due to human activities and their contributions to climate change. Naturally occurring changes in climate have been implicated but appear to be insignificant, especially relative to the human-induced onset of global warming, which has drastically altered natural processes, to the point that they will be forever linked to our existence.

The human impact on biodiversity first expressed itself a few millennia back in Earth's time. Between 10,000 and 50,000 years ago, our ancient ancestors appear to have played a central role in the disappearance of large mammals. The woolly mammoth disappeared about 3,700 years ago, coincident with a period of climatic warming, which reduced the animal's habitat. The mammoth, however, had managed to survive such warming periods prior to the appearance of human hunters.

With the emergence of the first organized human civilizations, human activities became focused on manipulating the immediate environment. Rather than only hunt large mammals or gather plants, humans learned to bring certain species under their control, eventually producing populations of domesticated livestock and crops. Domesticated species required space, and as villages grew into cities, more land and shelter were needed. As civilizations have grown and expanded throughout history, the continuity of Earth's environment has become increasingly disrupted. Animals that could once travel across wide expanses of forests and grasslands are now confined to fragmented habitats surrounded by walls of humanity, their populations stunted and their survival threatened.

Up until the latter half of the twentieth century, most knowledge of human-induced extinction focused on the loss of individual species. Among the extinct creatures whose stories have been repeatedly told are the dodo, which disappeared from Mauritius in the late seventeenth century, having been hunted to its end, and the passenger pigeon, the last wild one having been shot in 1900 and the last captive bird having died in 1914. Over the last several centuries, many other species have disappeared from Earth, far more than could have been lost by natural law alone. According to the International Union for Conservation of Nature and Natural Resources (IUCN), from the 1500s to 2009 some 875 species, including the dodo and passenger pigeon, went extinct. Approximately 114 of these were plants, having gone extinct entirely or extinct in the wild. Examples of recent, known plant extinctions include the St. Helena olive in 2003; the liverwort *Radula visiniaca* in 2000; a member of the amaranth family,

Blutaparon rigidum, in 1999; the St. Helena redwood tree in 1998; and a member of the Chrysobalanceae family, *Licania caldasiana*, in 1997. Most instances of modern plant extinction are related to overharvesting, overexploitation, and habitat loss. Some plants have not been found in the wild for many years and now are believed to be on the edge of extinction. They have not yet been pronounced extinct, though the encroachment of humans on their habitats is hastily unraveling the threads of hope for many of them.

The stories of extinct plants are rarely told. In most instances, so little is known about the plants themselves that there is not much to say. For recently extinct plants, however, this is changing. Research on species that disappeared within the last decade tells us that all perished under the overwhelming forces of human existence.

The downfall of plants endemic to St. Helena Island, a 47-square-mile area in the South Atlantic, began with the appearance of human settlers in the sixteenth and seventeenth centuries. The island was settled despite its geographical isolation—several thousand miles from South America and more than one thousand miles from Africa—and its rocky, volcanic ridges, deep gorges, and small size. The harvesting of trees for firewood and construction materials, and the clearing of land for plantations and grazing pigs and goats, reduced and fragmented the island's natural habitats. In this broken state, only small populations of wild species could survive.

Over the course of geologic time, the island's isolation has resulted in the evolution of unique species. Included among its endemic animals is the critically endangered St. Helena plover, and among its plants are the St. Helena olive tree (extinct), the black cabbage tree, St. Helena ebony, St. Helena boneseed, and the entire *Commidendrum* genus, which contains the bastard, false, and St. Helena gumwoods and a scrubwood species. It was inevitable, following human settlement and the ensuing events, that the numbers of endemic plants and animals on the island would decrease.

The St. Helena olive tree was among the species presumed extinct in the centuries following human settlement. In 1977, however, a single plant was found in the wild near the island's highest mountain, Diana Peak, and for a moment, there was a glimmer of hope that *Nesiota elliptica* could be saved. But this last plant was damaged by fungal disease, and it eventually died in the mid-1990s. The only successfully cultivated plant perished in 2003.

Conservationists on St. Helena are working to prevent the loss of other endemic plants that could share the same unfortunate fate as the olive. At the end of the nineteenth century, the bastard gumwood, which was described between 1806 and 1810 by botanist William John Burchell, who presumably named the plant based on a name given it by St. Helena locals, was believed to have gone extinct on the island. *Commidendrum rotundifolium* had sustained numerous insults, including having its bark gnawed off by goats brought to the island by settlers. In the early 1700s, in order to mediate the damage inflicted on the island's limited wooded areas, a planting law was established. This ensured that gumwoods and other trees used for timber were replanted by settlers. However, the law was successful for only a short while, and ultimately gumwoods were harvested to the brink of extinction. As with the olive tree, for the greater part of the twentieth century many believed that the bastard gumwood was gone forever. But in 1982, much to the astonishment of local botanists, a single tree was found. Though it died several years later, seeds that had been collected from it were successfully germinated. Many of the cultivated successors, however, have died, and by 2009, only one plant remained. A focused recovery effort led to the successful cultivation of new seedlings in the following years. However, it was suspected that most of the seedlings were infertile and would not grow when planted.

Because *Commidendrum* is endemic to St. Helena, there is additional concern over the fate of the whole group. False gumwood, *C. spurium*, is listed as critically endangered, with only eight plants left in the wild. *C. spurium*'s initial decline was due to the same factors that affected the olive and the bastard gumwood. Conservation of both gumwoods has been a particularly frustrating endeavor. The plants' genetic protection against inbreeding, a phenomenon known as self-incompatibility, has thwarted their cultivation. In the case of the bastard and false gumwood trees, self-incompatibility rests in an individual plant's ovules and pollen. Such protective mechanisms are vital to maintaining genetic diversity throughout a species' evolution, essentially preventing a plant from fertilizing itself. In many plants, self-incompatibility is mediated by their genetic codes, which contain a sort of inbreeding alarm known as the "*S* gene." If the plant's pollen grains come into contact with its own ovules, the presence of the *S* gene causes reproduction to abort, and thus a new plant does not develop. As a consequence of self-incompatibility, although hybrid bastard/false gumwoods can be cultivated, they cannot be successfully inbred to propagate new, purely bastard gumwood or purely false gumwood plants. Many other plants find themselves in this same situation, and identifying and protecting these species before they are reduced to a handful of living specimens is crucial to preventing their extinction.

Understanding how our actions, as well as natural factors, result in the endangerment of species forms the basis for developing effective and long-lived solutions for their protection. The ways in which we depend on nature, and on biodiversity in particular, are often overlooked. Simple things, such as the ability of nature to provide us with clean water and rich soil, are afterthoughts for many. But these basic environmental elements and processes support our survival, and because of this they are described in scientific and economic terms as ecosystem services. Food, water, wood, control of erosion and floods, waste processing, nutrient cycling, regulation of climate and air quality, medicines, aesthetic value, ecotourism, cultural heritage, and space for recreation are examples of ecosystem services.

But our modern lifestyles have severely impaired the natural functioning of ecosystems. Our freshwater supplies have become increasingly polluted, topsoil has eroded away from years of unsustainable farming, and old-growth forests have been cut to the ground, causing long-term ecosystem damage. In the coming years, nature's services may no longer be able to supply our demands. We are beginning to confront many of these issues and are coming to terms with the fact that we are consumers, using and often laying to waste everything from trees to oil to fish. But there are still many people who believe that these resources will never be depleted—or at least not in our lifetimes.

The belief that nature is here to serve us and will last forever has encouraged people to inflict many harms on the environment. We have cut down forests to build our homes, have blown the tops off of mountains to mine the minerals they contain, and have carved

The clean air, clean water, and aesthetic solace provided by nature are examples of ecosystem services. These services are readily apparent in places like the juncture of Presque Isle River and Lake Superior, the cleanest of the Great Lakes, in Porcupine Mountains Wilderness State Park, Michigan. (Photo credit: Jeremy D. Rogers)

out open spaces where our cities and towns can exist and swell. More than half the world's population now lives in sprawling urban areas that have dismantled nature. We have shunted water from wetlands and lakes to irrigate our crops, and we have introduced species of plants and animals into habitats where they do not occur naturally, with the result that they choke out native life, often triggering a cascade of ecological devastation.

Not surprisingly, many human activities have left plants and animals susceptible to extinction. Global warming, desertification, continued use of nonrenewable energy resources, excessive generation of waste, and unsustainable food production are major threats to the long-term survival of nature and humanity. Within individual countries, socioeconomic factors, such as low gross national product, high carbon emissions, dense populations, and high birth rates, have been associated with increases in the susceptibility of species to extinction. Plants and animals and natural landscapes are irreplaceable. With their disappearance, the beauty and majesty of Earth's most diverse and scenic places is diminished. And with the loss of each species as a result of our actions, we suffer, especially in spirit.

To prevent the disappearance of natural lands and wildlife, we need first to identify which places and species are at greatest risk, and then we need to figure out why. Biologists, ecologists, climate scientists, and a vast number of other researchers have set to work in places around the world to assess the status of existing species and the health of their habitats. In the process, they are also assessing the extent to which these creatures and lands are placed at risk from our activities. The IUCN has played a leading role in compiling this information. The organization has established a list of criteria for the evaluation of species to determine whether or not a plant or an animal should be added to their Red List of Threatened Species, which identifies species at greatest risk of extinction. Criteria evaluated include population size, geographical or ecological location, and length of time between generations. Other factors, such as amount and speed of decline in population number or reduction in geographical distribution, are also assessed.

But figuring out which species are at risk and how quickly they are disappearing has been difficult. Extinction is a natural process, and before scientists can determine just how extensive are our impacts, they must first have knowledge of existing species and natural patterns and rates of extinction. Researchers have described just short of 2 million of Earth's species. By far, the greatest number and greatest diversity occur in rain forests and similar dense habitats, and for reasons such as this, not all places in the world share the same pattern and rate of extinction. Species loss appears to favor places such as tropical rain forests, because these areas have high concentrations of endemics. The localization and concentration of endemic species in threatened regions are factors that are essential in understanding why some places appear to be extinction nuclei, with high rates of loss, and why other places, with similar geographical characteristics but with fewer endemics, experience a much different rate of extinction. Based on this, and on what has been learned from fossil studies and investigations of the evolutionary history of various species, scientists have determined that the natural rate of extinction is 1 species lost in every 1- to 10-year period. Human factors have increased this rate by between 1,000 and 10,000 times, with the most biologically rich and diverse places under the most extreme human pressures representing the high end of the spectrum.

When we think about endangered species, most people think first of animals, not plants. The loss of animals stirs up complex emotions. It upsets us, sometimes angers us. But the disappearance of plants and small animals, such as insects and invertebrates, living organisms buried in the depths of forests and other habitats where they exist out of our view, is a silent and invisible process.

About 380,000 species of plants have been identified, a figure that is increasing annually. Of all life-forms, plants are the most threatened by human activities. Of the flora found in tropical regions of the world, for example, between 19 and 46 percent may be at risk of extinction. In 2011 researchers from Kew, the Natural History Museum, London, and IUCN reported that globally about 20 percent of all plant species are at risk. Their findings were based on an assessment of 7,000 species representing different groups of plants in different parts of the world.

Botanists suspect that there are many plants that have not been discovered. Many of these species are believed to be lying deep within the world's rain forests, submerged in aquatic habitats, or buried beneath the soil in places such as deserts and grasslands. The notion that numerous species are yet to be described is thought to be the case not only for plants, but for all other groups of life too. In fact, it is believed that only a fraction of the total number of species currently dwelling on our planet have been described. Some scientists have proposed that another 3 million to 50 million species are still to be discovered. This would mean that Earth might hold anywhere between 5 million and tens of millions of different species. The possibility that so much life remains to be found has added a new dimension to the necessity and urgency of conservation efforts.

The loss of biodiversity, including the loss of plants and the insects that pollinate and coexist with them, represents a loss of potential medicines. And, beyond this, the fragmentation of indigenous cultures, in which are rooted some of the most ancient medicinal practices and an extensive empirical knowledge of plant medicines, is symbolic of a disappearance of information on which modern pharmacology and botany depend. Thus, there are reasons, some obvious and others more complicated, that explain the vital importance of cataloging Earth's life and identifying at-risk species. These efforts are now in a race against time.

The exploitation of trees and the heartwood they possess for economic gains, ranging from construction to agriculture, has been particularly devastating in the Americas, Africa, and Asia. Deforestation in the form of clear-cutting and even selective harvesting can come with serious consequences for local habitats, increasing risk of fire, causing changes in local climate, and reducing Earth's overall carbon-dioxide-absorbing capacity. In the 1980s and 1990s deforestation became such a huge problem that intense efforts were made to increase awareness of the environmental consequences and to reduce clear-cutting and other unsustainable practices. In the early twenty-first century, the loss of the world's forests continued at a rapid pace, with about 50,190 square miles (32,121,600 acres) of land deforested or lost to natural factors annually between 2000 and 2010. Because replanting efforts were stepped up, the actual net loss of forests each year during that time was about 20,080 square miles (12,851,200 acres).

A significant percentage of Earth's plant species reside in tropical rain forests, and because these areas support the survival of countless animals, extinction on all levels is a concern. In Amazonia, grasslands have taken over deforested areas, causing increases in local temperatures and decreases in precipitation, changes that have in turn given rise to longer dry seasons, leaving the grasslands and bordering rain forests susceptible to wildfire.

The disruption of habitat through the creation of clear spaces causes cascading effects within ecological communities. The loss of whole stands of trees not only affects the population dynamics of local animals and plants but also can cause severe disruption of migratory paths followed by birds and other species. Species whose habitats have experienced sudden change are forced to disperse to neighboring areas. In the case of certain insects and rodents, such changes can lead to infestation, defoliation of plants, and disease.

On the level that it has been taking place, deforestation also has been a factor in global climate change. The leaves, roots, and other parts of plants absorb carbon dioxide from the atmosphere, serving as carbon reservoirs, with all the world's plants combined storing nearly 290 gigatons of carbon (a gigaton is 1 billion metric tons).

When trees are cut down, they release carbon back into the atmosphere. The soil surrounding the harvested trees also gives off carbon. In tropical regions of the world, these processes, as a result of deforestation, accounted for an estimated 10 to 25 percent of carbon emissions worldwide in the 1990s. In the first decade of the twenty-first century, an estimated 0.5 gigaton of carbon was released each year from the carbon stores of forests as a result of loss of forested land.

Central and South America are known for their rain forests and diversity of plants, insects, and other animals. Because of their high degree of endemism and the perpetual threat of deforestation, they are recognized as biodiversity hotspots. The Mesoamerican hotspot in Central America, which extends from the Panama Canal northward into central Mexico, is one of these regions. The area of the hotspot originally consisted of about 436,300 square miles (279,232,000 acres) of vegetation. Its varied habitats, encompassing dry subtropical areas, mountains, and wet forests, support a wide range of plants, from cacti to juniper, pines, and other conifers to hardwoods such as mahogany. As many as 24,000 different species of plants have been described throughout the region, between 2,900 and 5,000 of which are believed to be endemic. But only about one-fifth of the original vegetation found within the area of the Mesoamerican hotspot is left standing today, and at least several plants have gone extinct in the wild.

Several endemic plant groups are known from a single species, making them particularly vulnerable to extinction. Among the region's casualties was the root-spine palm, which was native to Honduras. Now extinct in the wild, *Cryosophila williamsii* once thrived in the alkaline soils in the low elevations of the Honduran rain forests. It was used by locals for roofing material and as a source of food and medicine. It owes its common name to the sharp, branching spines that originated in the plant's roots and grew from its stems. All members of *Cryosophila*, which contains nine species, share this spiny trait, and all are confined to a narrow strip of geographical real estate extending from northern Colombia to western Mexico.

The *Cryosophila* genus is barely hanging on. Deforestation and the expansion of agriculture and human populations, which caused root-spine's extinction in the wild, have impacted other *Cryosophila*. In Costa Rica, *C. grayumii* and *C. cookii* are critically endangered, with the greatest threat to their survival being habitat loss. Fewer than 100 individual *C. cookii* plants now exist in the rain forests hugging the country's Atlantic coast. The limitations imposed on its range of pollination by agricultural invasion and deforestation are significant hurdles to its recovery. *C. grayumii* faces similar problems. Its several known subpopulations are concentrated primarily in the lowland rain forests on the country's Pacific stretch. A sister species found in Panama, *C. bartlettii*, is also listed as endangered. The fate of this plant seems to have been determined by the construction of Lago Alajuela (Madden Lake) in the 1930s. Alajuela is an artificial lake located at the western edge of Parque Nacional Chagres, approximately halfway between Panama City and Colon. The lake was formed from the creation of Madden Dam, and since its formation, it has wiped out an important expanse of *C. bartlettii*'s native habitat and has brought with it increased human activity. Only a small portion of the Mesoamerican hotspot's total area is under some form of protection. Most of this protected land falls within the borders of Costa Rica and Belize.

The region's plants and animals remain at risk of extinction from deforestation. Land has been cleared in Central America to create space to graze livestock and plant crops such as coffee and bananas. The twentieth-century invention of mechanized logging gave timber workers the ability to harvest large areas of virgin rain forest quickly. The annual rate of deforestation in Central America in the 1990s was between 0.8 and 1.5 percent, which actually was low compared to some places in the Amazon River basin in Brazil and in Myanmar (Burma) and Sumatra in Southeast Asia. The state of Acre in southwestern Brazil, home to much of the forested portion of the Amazon basin, lost 4.4 percent of forest cover annually in the 1990s. During that same period, Myanmar lost at least 3 percent annually, and Sumatra lost between 3 and nearly 6 percent. While trees can be replanted, the diversity of life that exists in the world's untouched rain forests cannot be replicated on a tree farm.

The idea for identifying and defining areas of the world that are rich in biological diversity and in need of research and conservation took flight in the 1980s with biodiversity specialist Norman Myers. In the following decades, Myers refined the qualification criteria, such that

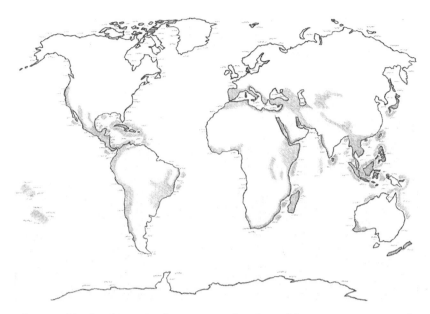

The world's biodiversity hotspots and vulnerable warm-water coral reefs (shaded regions). (Source data: Conservation International and UNEP)

now there are nearly three dozen regions of the world that are labeled as biodiversity hotspots. All are defined by the loss of more than two-thirds of their original habitat and by high levels of endemism—in fact, under strict definition, all hotspots contain less than 30 percent of their original habitat and are home to a minimum of 1,500 known species of native vascular plants. The combination of substantial endemism and severe habitat reduction means that hotspots are at exceptionally high risk of biodiversity loss and species collapse. At one time, the first 25 hotspots that were formally recognized blanketed nearly 12 percent of the planet's entire land area. By 2010, however, these 25 areas covered less than 1.4 percent. Amazingly, even with this significant reduction, these regions still collectively hold 44 percent of endemic plant species.

While deforestation has played a major role in reducing the biodiversity and area of regions now listed as hotspots, other human activities have also contributed to their destruction, and in ways more permanent. Estimates of population size, food production, and water needs in the coming decades paint a telling picture of modern human society. We are headed down a path that diverges sharply from nature, and we are being chased by a ferocious, uncontrollable rate of population growth. There are more than 6.7 billion people living in the world. An estimated 3 billion more will be around in 2050. This unprecedented rate of growth has already had substantial effects on agriculture and natural habitats.

Headed in the opposite direction of population data is the percentage of Earth's surface that remains suitable for agricultural use. About 11 percent of the planet's total land area can be farmed, and this percentage is decreasing on a yearly basis. In some places, arable land has been abandoned, either because of climate change that has caused local drought, forcing increased reliance on unsustainable irrigation, or because of overgrazing and other anthropogenic factors that have led to desertification of the land. Within countries, reduced agricultural output results in an increased dependency on imported grains. By 2050, Earth will have barely enough water and agriculturally useful land area to support the global population. And those calculations ignore the direct consequences to wildlife and ecosystems produced by such an abundance of humans.

Ways to fulfill food demands created by increasing numbers of people need to be considered carefully. For example, methods of mass farming of animals, namely in industrialized countries, have increased the amount of meat and dairy products on the shelves of grocery stores, but a tremendous amount of grain is wasted on feeding these animals. So much grain, in fact, that one ecologist claimed that about 840 million people worldwide could live off all the grain that had been fed to livestock in the late 1990s in the United States alone. Considering that US livestock populations have expanded since, this figure has likely increased by now. Livestock also emit very large quantities of methane, which is many times more effective at trapping heat in the atmosphere than carbon dioxide, another gas they produce. Globally, farmed animals actually produce amounts of greenhouse gases similar to or greater than automobiles. Through overgrazing, these animals have also contributed to the destruction of wildlife habitats, pushing some species into extinction.

Most biodiversity hotspots are affected to some degree by overpopulation. The Himalayan mountain system, which spreads across the northern edge of India, encompasses a total area of approximately

231,660 square miles (more than 148 million acres). One would think that an area so remote and rugged would have a relatively sparse population, and of course compared with the rest of India, it does. But within the region of the Himalayas, there are broad variations in population density. The lowest, about 13 persons per about 0.4 square mile, is in Arunachal Pradesh, which covers slightly more than 32,330 square miles (20,691,200 acres) in the eastern slopes. The highest, at 510 persons per roughly 0.4 square mile, is the Darjeeling district in the West Bengal Hills, which occupies an area of about 1,216 square miles (778,240 acres).

At one time, a major portion of the Himalayas was covered with vegetation, but about 75 percent of the native plant life has disappeared, stripped away by overgrazing, deforestation, and crop farming. The overharvesting of medicinal species has also been a significant factor. In Nepal, 2.6 million people harvest medicinal plants from the mountains. This intense collecting activity, in addition to habitat loss, has threatened multiple plant species. Among the highest priority for conservation are spikenard, *Neopicrorhiza scrophulariiflora*, and *Angelica glauca*, which experience excessive levels of harvesting specifically because of their medicinal properties.

There is wide variation in harvesting practices for species of Himalayan medicinal plants. Methods employed by trained practitioners of Tibetan medicine who gather the plants for use within their local communities are far less damaging than the methods used by persons from commercial entities, who distribute the plants to places outside the local region. Local practitioners place much greater emphasis on sustainable harvest, whereas commercial harvesters view the plants largely in terms of monetary value and are less selective about which plants they collect. The greatest amount of commercial harvesting of spikenard and *N. scrophulariiflora* takes place in regions bordering Himalayan national parks, resulting in drastic reductions of the plants in these regions. The sustainable practices of local practitioners, who are permitted to harvest the plant both inside and outside park boundaries, have played a vital role in demonstrating how wild plant populations can be maintained while also fulfilling roles within human communities.

When it comes to population density and habitat destruction, however, the Himalayas are surpassed by the Indo-Burma hotspot, which covers the far eastern portion of India and encompasses Myanmar, Thailand, Laos, Cambodia, Vietnam, and the southern rim of China. Indo-Burma is home to about 7,000 species of endemic plants and close to 2 percent of all the world's vertebrate species. Its wealth of diversity lies within an area of about 810,800 square miles (almost 520 million acres), though today only 5 percent of it remains untouched. Just about everything, from too many humans to deforestation, mining, and wetlands draining, has given rise to the desolate state in which the hotspot currently finds itself.

In 2000 scientists reported that India's major hotspot, the Western Ghats and Sri Lanka region, retained only about 7 percent of its original vegetation, a figure that has likely declined further in the last several years. The Western Ghats, which makes up the northern section of the hotspot, consists of a narrow mountainous strip of lush forest running north to south along the country's western margin, skirting the edge of the Arabian Sea. It consists of a series of mountain peaks, the highest of which tops out at nearly 8,860 feet. These peaks rise up between plateaus, upon which open grasslands spread across gently undulating landscapes. The variability in rainfall on the western and eastern slopes and the intricate and knotted terrain found in the interior mountains give rise to a broad range of vegetation, including tropical and deciduous forests.

At one time, India and Sri Lanka were connected by what is known as Adam's Bridge, a series of shoals that are now remnants of a solid link between the two places. Some 2,180 endemic species of plants may still be found in these areas. Many inhabit the higher reaches of mountainous and plateau terrain and possess medicinal properties.

In India's neighbor to the north, overpopulation is one among many different anthropogenic factors that have affected biodiversity and the general health of the environment. China covers a land area that is three times larger than India, but large swaths of the country, because of their general lack of agricultural resources, are unsettled. Many people have taken up residence in areas with favorable climates and soils that are easily farmed. One such area is the Loess Plateau, arching over the country's central expanse.

The plateau is covered by an abundance of fine-grained silt sediment, or at least it was. Loess is very fertile and productive for farming, but it is highly susceptible to erosion by wind and water, a feature made even more conspicuous following human manipulation of the land. The landscape of the Loess has been dramatically altered by centuries of human intervention. Villages in the region were often formed on hilltops, sites where crops could be grown easily. As local populations grew, farmers expanded into the surrounding sloping lands, carving up the hills, stripping away soil layers millions of years old. Terracing on gently sloping lands improved rainwater catch. But this strategy proved wholly ineffectual on steep slopes, and decades of failed attempts to tame the abrupt inclines only worsened erosion and further dismantled natural habitat. The devastation and consequent restructuring of the Loess Plateau are signs of China's ongoing population and agricultural crisis. There are simply too many people, and this has led to the too-rapid consumption of natural resources, especially water. Only a fraction of the plateau's native plants and animals still exist there today.

A short ways south of the Loess the mighty Yangtze River surges across the plains of China, rushing forward uncontrollably, pressing hard on the recently constructed Three Gorges Dam. The behemoth concrete barrier was built to harness the energy of the Yangtze, but it so far has proved to be an extraordinarily difficult industrial project. Besides the displacement of more than 1 million people, the backup in water flow has drowned waste sites, factories, and houses, leading to severe water pollution. The dam was opened in 2006, when the main wall was finished, and since then, the consequences of the project have trickled steadily and uneasily downstream. The coastal lands of the east, particularly at Shanghai, and even as far north as Beijing are subject to the disastrous environmental fallout. Fish caught from the East China Sea form a major food source. Already suffering significant declines from overfishing, the accumulation of silt and waste where the river meets the sea is expected to cause major die-offs of the sea's fish.

Much of the western portion of the Yangtze lies within the Mountains of Southwest China hotspot, which has been heavily affected by the disruption of water flow as a result of Three Gorges. Although the actual human population density and overall area of this hotspot are much lower than its two neighbors, the Himalayas to the west and Indo-Burma to the south, the fallout from events set in motion by the crowding of humanity into the surrounding regions has taken its toll on this more remote land. The South-Central China Region described by Norman Myers in 2000 housed an estimated 3,500 endemic plant species. But less than 8 percent of the original vegetation-inhabited land area remains.

The variations in elevation and surface features, combined with dramatic differences in climate from north to south and west to east, make the Mountains of Southwest China hotspot one of the most intricate regions identified to date. An estimated 2,000 species of medicinal plants are found there, and living in the wild of the forests and on the slopes of the mountains are several high-profile threatened animals, including the giant panda, the golden monkey, and the endangered snow leopard. Among the region's many critically endangered plants are multiple species of orchids.

Within the region, wild populations of the orchid *Dendrobium officinale* have collapsed. The species' downfall points to two major factors contributing to the overall loss of plant biodiversity in southwestern China: the fragmentation of habitat and the overharvesting of medicinal plants. *D. officinale* has been used in traditional Chinese medicine for many centuries, but the breaking up of habitat associated with the escalation of deforestation beginning in the twentieth century left small populations isolated from one another, stranded on islands of green in a deranged sea of human development. For any species, living in such a severed state, with few individuals scattered here and there, is often impossible to overcome. For animals, such small populations may eventually resort to inbreeding in order to survive. Inbreeding reduces genetic diversity in the population, and low numbers and reduced genetic diversity leave populations susceptible to disease. In plants, only those species that can reproduce by self-pollination are able to inbreed, and there are few capable of doing so. Rather, the majority depend on pollinators or seed carriers to bridge the habitat gap. While maintaining genetic diversity is extraordinarily important for plants, most species affected by habitat fragmentation will likely struggle to survive because of the basic inability to produce new generations. Normally minor climatic variations suddenly become magnified. An unusually dry or wet year can potentially decimate a fragmented population.

D. officinale's situation worsened substantially when increasing commercial demands for its medicinal use led to the overharvesting

of remaining wild populations. Its subpopulations continue to con-
tract. The species is being shoved toward the point of no return,
sliding closer to the edge. It thrived prior to the twentieth century,
before the introduction of modern agricultural technologies and the
problem of overpopulation.

The Mountains of Southwest China hotspot has encountered
other threats too, including the overgrazing of livestock, goats especially, which are universally known to eat anything and everything.
The hunting of animals whose parts are highly valued in Chinese
medicine has also pushed species near to extinction. Mining and the
construction of more dams on rivers other than the Yangtze have
flipped the natural ecosystems of this region upside down.

There are many other stories about vanishing species in China,
and each can teach us important lessons about nature and the harm
we inflict on it. In the case of ginkgo, we discover that even the
strongest and most evolutionarily exquisite species are, in their wild
habitats, still no match for humans. The story of ginkgo began long
ago, several thousand years ago, in fact, when it became cherished in
traditional Chinese medicine. *Ginkgo biloba* trees are gymnosperms
with an impressive evolutionary history. The *Ginkgo* genus represents
the oldest group of living trees on Earth—the first of their kind
appeared between 200 million and 170 million years ago, during the
Jurassic period. *G. biloba* later evolved within this clade, sometime
around 100 million years ago.

Ginkgos are resilient. They survived the End Cretaceous extinction, which did away with many families and genera of land plants,
and they are exceptionally adaptable to different soils and climates.
In the modern era, however, the survival of ginkgo trees has become
dependent on humans. The once diverse *Ginkgo* group is today represented by a single species, *G. biloba*, which actually may be a cultivated version, rather than the species that originated in the wild. If
this is the case, then the species with which many of us are familiar
is genetically distinct from its wild ancestor.

Deforestation and other factors whittled down wild populations
of ginkgo, and despite their popularity in sites of deliberate cultivation, only a few, possibly wild specimens have been found. These
vaguely wild trees grow on Mount Tianmu, in the eastern region
of China, and are found near Taoist and Buddhist temples, where

they are treated as sacred plants. Because of the uncertainties surrounding the existence of wild ginkgo trees, and because so few plants remain outside of cultivation venues, the species is listed as endangered by the IUCN.

The cultivation of ginkgo in China began about 2,000 years ago in the mountains and plains along the Yangtze. The nuts, which are edible, and the leaves, which possess medicinal properties, are the most valued parts of the plant. The leaves contain a variety of bioactive substances that serve as natural plant defenses against insects, disease, and ultraviolet radiation. The presence of these substances likely provided an evolutionary advantage to ginkgo trees, enabling them to overcome many natural obstacles to which other plants succumbed.

Many of the compounds present in the plant's leaves have been isolated and characterized, and most over-the-counter ginkgo preparations currently available contain a standardized extract known as EGB761, which is made from dried leaves. This extract contains some of the plant's most potent antioxidant compounds, scavengers of harmful free radicals that can cause damage to cells and DNA. Free radicals play a role in diseases such as cancer and in the overall aging process. At specific concentrations, the antioxidants in EGB761 can protect cells from radicals, which has important implications for the treatment of cardiovascular diseases such as atherosclerosis, in which oxidative damage to the lining of blood vessels enables fats and cholesterol to accumulate and eventually impede blood flow through the vessel, leading to stroke or heart attack.

There also has been significant interest in the detection of ginkgo compounds capable of protecting against Alzheimer disease. Chinese practitioners have used ginkgo for centuries as a sort of brain stimulant. But despite extensive research into various compounds and mixtures of compounds, it remains unclear whether ginkgo is medically beneficial in staving off Alzheimer or other forms of dementia. There are many other examples of conditions for which ginkgo is used as a traditional herbal remedy, but for which substantiating scientific evidence is sparse.

The tree also produces several compounds that are toxic in humans, including ginkgotoxin (4-O-methylpyridoxine), which causes "gin-nan" poisoning. When ginkgo seeds are consumed in excess, poisoning may ensue, causing seizures, paralysis, and death in severe cases. It is these types of defenses that facilitated the ginkgo's survival in the wild for millions of years. Humans have essentially uprooted the tree from the wild, and after thousands of years of cultivation it is likely that this human factor has changed the plant's genetic history, with the consequence that we and future generations to come may never know the wild species.

The story of ginseng shares threads similar to those of ginkgo. The Asian *Panax ginseng* is known in Chinese as *jen shen*, or "man plant," owing to the shape of its roots, which resemble the human body. Another medicinally useful species is *P. quinquefolius*, which is native to Canada and the United States. Asian ginseng's tortuously gnarled root is the part most valuable to medicine. Its genus name, *Panax*, is derived from the Greek word for panacea, which reflects the all-healing powers subsumed by the species. Included in its wide spectrum of traditional uses are providing a sense of well-being, releasing stress, and improving strength. In traditional Chinese medicine, ginseng is used to revitalize the life force known as *qi*. Representing vital energy and its equilibrium within the human body, *qi* is one of the two major principles underlying the practice of traditional Chinese medicine. The other, the five elements (fire, earth, metal, water, and wood), is also believed to be restored by ginseng.

Asian ginseng is native to the Changbai Mountains, where the Manchurian Plain in China ends and the mountainous depths of North Korea begin. Because the variety is more potent than North American ginseng, it is more highly sought after and as a consequence has nearly disappeared in the wild. At one time, it grew plentifully beneath conifers and deciduous trees. But its slow development, between five and seven years to reach full maturity, and its habitat requirements have made it exceptionally vulnerable to the challenges presented by human exploitation. Overharvesting and harvesting of the roots before a plant reaches reproductive maturity, before it can produce a new generation, in addition to deforestation, have driven the plant to the brink of extinction.

Global demand for ginseng is unbelievably high, with hundreds of tons of the North American variety, despite its inferior qualities, exported from the Americas to China each year to fill the country's demand. All Asian ginseng grown for medicinal purposes is now cultivated by farmers in China, Korea, Japan, and Russia. But because this is still not enough to fill Asia's demand, the North American plant must be produced in superquantities. Similar to Asian ginseng, to prevent the extinction of the wild North American variety, which has become increasingly rare in its native habitat, the plant must be cultivated.

The discovery of bioactive compounds in ginseng has proven the plant's value to modern medicine and provided substances suitable for the generation of new synthetic and semi-synthetic drugs. The development of such agents is expected to relieve the devastation of overharvesting of ginseng, buying time for wild populations in both China and North America to recover.

Although overpopulation and its resulting environmental destruction seems a universal theme of biodiversity hotspots, there are threatened regions characterized by very low population densities. Among these are the Southwest Australia and the Succulent Karoo hotspots. The former is sometimes referred to as the Southwest Australia Ecoregion, which is actually larger than the hotspot alone. It consists of eucalyptus woodlands, saline wetlands, shrublands, and aquatic habitats. Falling within this region is the Southwest Botanical Province, an area populated by a diverse collection of plants that was surveyed in 1980 by ecologist John Stanley Beard. Beard, whose work involved in part an assessment of flora found in the whole of western Australia, was one of the first to provide detailed information on the southwestern region's shrubland ecosystem.

The Southwest Australia hotspot is known primarily for its vascular plants, having somewhere between 5,570 and 6,750 different species, more than half of which are endemic. Many of these are representatives of very primitive species and manage to thrive in a climate of perpetual cycles of wet winters and dry summers. They also contend with a tessellation of soil types, many of which are wanting in nutrients.

At one time, a variety of eucalyptus trees, such as gimlet and York gum, reigned in dominating stands across large areas of Southwest Australia. Forests of karri prospered in the deficient soils, sometimes reaching 70 to 90 meters in height, placing them in the exclusive group of Earth's tallest trees. Along with species of *Banksia* wild-flowers, the Australian pitcher plant, and *Daviesia* shrubs, the euca-lyptus and all the rest of the diverse collection of flora in the region have evolved in an ecological system maintained by fire. These natural processes are key to the endemic species' growth, but their disruption, through the clearing of shrublands and woodlands for agriculture, the introduction of fertilizers, and various bauxite min-ing activities, has caused eucalyptus and the other endemic, iconic species of the region to begin fading into history.

The Succulent Karoo of South Africa has likewise sustained irre-versible damage, with mining, crop farming, and overgrazing by goats, sheep, and ostrich ranking among its greatest threats. Relative to hotspots in China, however, only very small areas of the Karoo's land are lost permanently. Rather, its concerns center around the future loss of biodiversity, primarily from agricultural expansion, mining, and climate change. Efforts to establish new protected areas, and to expand existing ones, are considered of utmost impor-tance. In the early 2000s, only about 3.5 percent of land within the Karoo biome was under protection, and more than 930 of its species were on the IUCN Red List.

The Succulent Karoo biodiversity hotspot of South Africa.

The Succulent Karoo has a rich variety of plants—more than 6,350 different species, more than 2,500 of which are endemic. Some of its flora overlap with areas of the Cape Floristic Region hotspot, which lies to the south and covers the tip of the African continent. The moderate and persistently arid climate of the Karoo, however, is distinct from that of the Cape, which enjoys a Mediterranean climate with wet/dry cycles from winter to summer. Both hotspots, however, are subject to severe losses in plant biodiversity in the coming years. With this loss comes the loss of genetic diversity. Similar to the way in which natural fire regulates ecological factors in southwestern Australia, there are important evolutionary and ecological processes that support the diversity of flora in South Africa. One of these processes is gene flow, which for many plants is influenced by seed dispersal and pollination. The ability to then spread and populate different areas is fed by adaptation and natural selection. Of particular interest for gene flow in the Karoo are succulents and geophytes. These two groups of plants make up the greatest number of the Karoo's endemic species and rely on unique adaptations, such as water-assisted seed dispersal over short distances or pollination by solitary bees. They also tend to thrive in quartz and gravel patches. Knowledge of these habitat needs assists in assessing the spatial and systematic requirements necessary for conservation and establishment of a national park in the Karoo.

Hotspots garner a significant amount of research and conservation attention. Such areas of condensed biodiversity enable more species to be protected per dollar than less biodiverse areas. But there are a variety of unique, lower-diversity habitats that are also in need of our attention, including grassland biomes such as the North American prairie. The US prairie is considered to be the country's most endangered ecosystem, with only 1 percent still left intact. Much of it was plundered by agricultural activities in the twentieth century, and today urban sprawl is pushing hard against the borders of the last little bit that remains.

Extinction has become universal across all Earth's ecosystems, even though it may be taking place at different rates and in variable patterns, making it difficult to detect or assess. In the modern era, habitats and the species they contain are at the greatest risk of extinction that they have ever known. The consequences on all fronts, from the aesthetic qualities of nature to the functioning of ecosystems to human health, are severe. In terms of medicine and our ability to fight disease, protecting Earth's plants and the compounds they produce has become a necessity.

6

Out of Nature

Ginkgo biloba.

NATURE IS UNDERGOING A STEADY process of deterioration. Its diversity has dwindled at a rate that has been nearly imperceptible to our species, a species now relatively out of touch with nature. But to those few among us who routinely experience and study nature, its peril and promise are unsettlingly conspicuous.

We can dwell on the peril in which nature now finds itself, but its promise—its potential, in its fully wild and natural state, to make our lives better—is especially intriguing. Discovering and rediscovering its wonders gives us an opportunity to revisit and explore the roots of our own species. And as recent ventures to investigate habitats and to catalog species have suggested, our return is fulfilling and necessary. Many things remain to be discovered in nature and to come out of nature, and so there is much, despite our struggling world, to be hopeful for.

Over the course of the last century, the health and survival of humanity have become increasingly dependent on conventional medicine and its drugs. On one end of the spectrum exist the drugs that relieve our aches and pains and that allow us to continue to function in our everyday lives. On the opposite end are vaccines, antibiotics, and antivirals, the only barriers separating us from potential global disasters caused by infectious disease.

Many of our most reliable agents have come out of nature, and for nearly the whole of human history, the practice of healing and curing disease has depended on the vitality of plants. Our ancestors were unaware of the causes of many of their ailments, but they relied on the prolific leaves and the stems, roots, and bark of plants to work healing miracles. Plants were steadfast in times when humans knew little about how the natural world worked, when supernatural capacities were ascribed to natural phenomena. During the Enlightenment, knowledge was acquired about Earth's many different types of creatures and how they function. Plants became known for photosynthesis and the conversion of carbon dioxide to oxygen, abilities generally unrelated to medicine. These discoveries astonished the scientific community. They united major components of Earth's atmosphere and the ability of animals to respire with the ability of plants to grow—a connection that few had postulated. The discoveries also came at the time when the exotic finds of plant hunters like Joseph Banks and Joseph Dalton Hooker renewed the sense of appreciation for the diversity and beauty of plants. Later, the influence of nitrogen on plant growth was discovered, catapulting crops into the spotlight of agriculture. Prairie grasses, plants of swamplands, and even desert cactus—plants that form the cornerstones of their ecosystems—lie defenseless against acres upon acres of fertilized and irrigated plants pampered and patronized by humans.

Plants today are broadly conceived of as domesticated things, mainly in the contexts of food and agriculture. The importance of Earth's vegetation to medicine and ecosystems has been eclipsed by the colossal scale of the subordination of plants to the basic level of monocultured food source. Our relationship with medicinal plants, on the other hand, is very different and quite unique. Unlike food plants, which our ancestors domesticated and therefore changed genetically, most medicinal plants are wild, unchanged by human hands. This, in addition to the many ancient stories of their ability to relieve symptoms of illness, gives these plants an unusual mystique.

As more has become known about plants, it has become clear that they are not only extraordinarily important to modern medicine, but against the backdrop of increasing knowledge of the world's climate and environment, many species are extremely fragile. The rarity of many medicinal species in the wild and the inability of rare species to survive under cultivation have defied the will and the strength of humans. Still, we find ourselves drawn to them, fascinated by them. In attempting to understand these species, the realization of their significance to modern science, to human survival, and to Earth's biodiversity has brought new meaning to medicine and conservation.

The majority of remedies used in systems of medicine that have been practiced since ancient times are plant based. Most of the people who rely on these medicines live in countries in Asia and Africa, where traditional medicine is practiced exclusively by about four-fifths of the population. Such remedies, however, are relevant not only in societies where the presence of conventional medicine is lacking. Between 70 and 80 percent of people in developed countries have used some type of complementary or alternative treatment. (Complementary medicine is used alongside conventional medicine, whereas alternative medicine is used in its place.) These forms of treatment include the use of over-the-counter products containing plant materials, as well as reliance on therapies such as massage and acupuncture. In conventional medicine, about one-quarter of prescription drugs contain bioactive chemicals derived from plants, and close to two-thirds of anticancer drugs are based on chemicals extracted from natural sources. There also are numerous drugs derived from compounds found in bacteria, molds, and other organisms.

The global value of plant-based medicines is most readily apparent in revenue generated from sales of over-the-counter herbal preparations. In 2005 more than $14 billion (USD) was spent on such remedies in China alone. In 2007 US citizens spent even more—nearly $15 billion—on over-the-counter natural products. Ginseng, ginkgo, echinacea, garlic, feverfew, saw palmetto, and kava kava ranked among the most frequently used herbal remedies. The amount of money spent on these products accounted for nearly 44 percent of all out-of-pocket expenses on complementary or alternative medicine in the United States in 2007. The surprising element of this is that the safety or efficacy of most nonprescription natural products is unproven. A small quantity, just micrograms, of a substance may be enough to exact a cure, but the average person buying an over-the-counter "natural" herbal remedy is often buying an extract that is too concentrated or that may not even be natural at all. Unapproved drugs facilitate self-medication, a dangerous and risky approach to healing for people unfamiliar with the physiological effects of the substances they use. Yet, when faced with the hurdles of privatized health care and the often exorbitant costs of conventional medicine, people would rather take matters into their own hands.

The decision to use traditional, complementary, or alternative treatments entails much more than dollars and cents. There is a greater, holistic sense of healing associated with these approaches, perhaps because they articulate the organic side of medicine. In the case of traditional practices, lifetimes of knowledge lie behind the art of mending human ailments. This does not mean, however, that we should consider every treatment applied in traditional systems of medicine to be all-healing. Some are not physiologically beneficial at all and benefits attributed to others may actually be the result of an associated placebo effect. Other concoctions may even harm the body. But within the extensive corpus of traditional knowledge there exists a wealth of plant remedies that scientists can turn to for investigation. Perhaps more importantly, there also endures a connection to the natural world, an aspect that is wanting in conventional medicine and many modern societies in general.

But conventional medicine produces proven, tested medicines, and it has saved more lives already than any other form of medicine in history. Much of its success has stemmed from the discovery of life-saving drugs, and in order for this to continue, in order for drugs new to medicine to emerge in the coming years, nature must become a part of modern discovery, with its role alongside synthetic discovery reestablished. In recent years, realizing that many species of plants may yet be discovered, scientists have found renewed hope for medicine and the healing of human disease. In the process, they have thrown into startling relief the need to protect the world's plants.

In nature, the multitude of chemicals in the leaves and stems of plants provide a concerted concoction of protective and nourishing substances, some of which are beneficial to humans and others of which are toxic. There is substantial variation in the quantities of chemicals within a plant's extracts, depending on ecological and geographical influences on growth and development and on medicinal preparation. The activity of plant chemicals can change dramatically with specimen collection and preparation. Thus, the isolation of a natural bioactive principle enables scientists to identify effects and toxicities and to use the compound as a lead for the development of synthetic derivatives. The latter is crucial for ensuring the reproducibility of effect and the generation of enough drug for commercialization, dissolving the need for constant collection and potential overharvesting.

Between 1950 and 1963, interest in plant-based natural-products drug discovery reached its zenith in the United States. During this time, although there was increased emphasis on synthetic chemistry, a number of researchers remained devoted to discovering and isolating new compounds from natural sources. In the 1950s, the National Cancer Institute (NCI) Developmental Therapeutics Program initiated a massive screening project to identify novel bioactive substances from plants. From 1960 to 1982, the investigation involved a variety of compounds, isolated from more than 35,000 different species of plants, with some 10 percent of these ultimately identified as sources for new anticancer agents. Progress was painstakingly slow, however, and as early as the mid-1970s, identifying chemicals from plant material was already challenged as an outmoded approach to drug discovery. The NCI screening project eventually died out, receiving a budget cut in 1981 that resulted in its termination, thereby ending one of the largest efforts at natural-products discovery in the United States. In the early 1990s, only seven anticancer drugs from the project had survived to late-stage clinical trials, persevering through the challenges presented by both synthetic chemistry and environmental concerns.

The extent of the problems facing plant-based natural-products discovery was daunting, and few scientists saw any promise in the field, particularly in the form of funding. Furthermore, with advances in biotechnology diminishing interests in investigation of the natural world, plant science and natural-products discovery seemed antiquated endeavors. In the 1980s, interested in capitalizing on the improvements in biotechnology, many universities transitioned their research goals toward technological advance, leaving academic labs focused on the investigation of nature and plant compounds to atrophy. This lasted well into the 1990s, improving marginally only recently, such that now a small proportion of researchers in the pharmaceutical industry and across academia share a renewed sense of importance in natural-products discovery.

If any refocusing of scientific attention on drug discovery in nature in recent years can be said to have happened, it has occurred almost exclusively as an improvement in scientists' knowledge of the chemical structures of bioactive compounds from plants and marine organisms. To expand this knowledge to applications in medicine, there also needs to be an accompanying increase in efforts to achieve a better understanding of how plants are used in traditional systems of medicine and to discover new bioactive compounds in plants. Much of this work, already under way, entails basic research, but a steady increase in the number of plant compounds that cross the threshold into clinical trials is needed as well, if entirely novel drugs are to continue to reach the market.

This book opened with the story of Taxol, a drug that is made semi-synthetically based on a compound originally discovered in the bark of Pacific yews. But Taxol is not alone in the realm of drugs that have come out of nature. Indeed, there have been many others, from antimalarial drugs such as artemisinin and quinine to the glaucoma drug pilocarpine, the anticancer agents vinblastine and vincristine, and pain relievers like morphine and aspirin. Each of these drugs has a story to tell, and though the characters change, the common thread perpetuates the importance of drug discovery to human health and, in turn, the importance of nature to medicine.

The Greek goddess Artemis was said to be the guardian of vegetation, wild animals, and women. *Artemisia*, in treating the menstrual disorders of women, was given the goddess's name. Today, the plants have been placed in the family Asteraceae, which contains about 23,000 to 24,000 species, and it is sweet wormwood in particular that has significance in modern medicine. The leaves of *Artemisia annua* were traditionally used to make qinghao, a tea with activity against hemorrhoids and fever. The plant's bioactive compound, qinghaosu, which knocks out all the parasites that cause malaria, is known commercially as artemisinin, an extraordinarily effective antimalarial drug that has few side effects in humans. When used together with other antimalarials in combination therapies, the risk for emergence of malaria parasite resistance is low.

Sweet wormwood is cultivated on a commercial level mainly in Vietnam, China, and Africa. Demand for the drug is very high, however, and artemisinin farms are struggling to sustain the necessary supply. While efforts are under way to increase agricultural production, part of a program headed by the World Health Organization (WHO) and the Roll-Back Malaria Partnership, scientists have been searching for ways to generate artemisinin in the laboratory. One method under investigation employs a genetically engineered strain of the yeast *Saccharomyces cerevisiae*. The yeast has been designed to contain sweet wormwood genes, enabling it to produce the precursor to artemisinin, a substance known as artemisinic acid, which can then be purified and formulated into artemisinin. The yeasts, however, are subject to cellular stress as a result of producing superquantities of the acid, an effect that has complicated the process of scaling up to commercial levels. Despite this, there is hope that microbial artemisinin will be able to relieve cultivation demand for sweet wormwood in the future.

Analyses of artemisinin's chemical structure led to the development of artesunate and artemether, semi-synthetic derivatives. These

Sweet wormwood (*Artemisia annua*), the source of the antimalarial drug artemisinin. (Photo credit: Jorge Ferreira)

compounds have been used to treat malaria, though because they are semi-synthetic, their generation requires artemisinin, which must be isolated from sweet wormwood. As with artemisinin itself, its derivatives are effective against drug-resistant strains of parasites and are used in combination with other antimalarials to prevent resistance. In Thailand and Cambodia, artemisinins were prescribed and administered in single-drug regimens, resulting in the development of resistance. While this has hampered their utility in those regions, they remain vital in the war against malaria in places such as Africa and Central and South America. Artemisinin can also block the replication of cancer cells, and its derivatives appear to possess similar activity, indicating that the entire family might also have a future in the treatment of malignant disease.

Artemisinin is remarkable for its recent discovery, having spun out of China's mid-twentieth century effort to turn its ancient herbal remedies into conventionalized medicines. A much longer and storied history, however, lies with quinine. Quinine was first isolated from the bark of *Cinchona officinalis* in 1820 by two French pharmacists, Joseph-Bienaimé Caventou and Pierre-Joseph Pelletier. Cinchona first appeared in Europe in the seventeenth century, under the names Jesuits' bark and Peruvian bark, having been brought there from South America. It is believed that sometime between the late 1500s and early 1600s, Jesuits living in Lima learned of a cinchona bark preparation from the Quechua people. Because malaria probably arrived in South America with the arrival of Spanish conquistadors, the natives could not have been using the preparation specifically for malaria. Nor were they using it to treat fevers from other causes, since cinchona bark is effective in reducing fever only from malaria. The generally accepted conclusion is that the natives used it as a sort of warming remedy in cool weather, since in low doses quinine causes an increase in body temperature. Quinine is an alkaloid, an organic compound made up of a ring structure containing nitrogen and at least one other kind of atom. All sorts of alkaloids have been discovered in plants and fungi, and most are biologically active in humans and other animals.

Although the use of quinine in the treatment of malaria declined starting in the 1940s, when other antimalarial agents were developed, today it is one of three medicines, the other two being the semi-synthetic artemisinins, that are effective for treating severe malaria. It is particularly valuable for instances in which parasites are resistant to pyrimethamine combinations and chloroquine antimalarials. Quinine is often used in combination with antibiotics such as tetracycline, clindamycin, and doxacycline. Since no plant material is required for its production, quinine manufacture never touches South American forests. And, like artemisinin, it too has resulted in synthetic derivative antimalarials, including mefloquine, chloroquine, and bulaquine. A drug related to quinine, known as quinidine, has also been isolated from cinchona. In addition to its use against malaria, quinidine is also valuable in the treatment of cardiac arrhythmia, or abnormal heartbeat. It too can be prepared synthetically.

Out of South America there have emerged many substances with applications in modern medicine. The plant sources of several, however, are in danger of extirpation. The alkaloid pilocarpine, found naturally in species of jaborandi shrubs of *Pilocarpus*, was first isolated from the leaves of *P. jaborandi* in 1875 and was later found in ten different Brazilian *Pilocarpus* species. But in only two of these, *P. jaborandi* and *P. microphyllus*, is the alkaloid produced at very high concentrations. The latter, which grows in the wild only in Paraguay and Brazil, has served as the primary source of pilocarpine since its discovery. The compound's medicinal value stems from its ability to diffuse into the eye by crossing the protective mucous boundary, the conjunctiva, that surrounds the eyeball. After successfully entering the eye, it stimulates receptors on the iris and causes contraction of the pupil. This effect facilitates the flow of fluid out of the eye and into special drainage channels, which can relieve pressure within the eye in persons with glaucoma. Pilocarpine also acts on receptors in the central and peripheral nervous systems, causing increased salivation and perspiration. For most people, either of these effects is undesired, but for those with dry mouth (xerostomia), the former is a godsend.

Jaborandi was used medicinally by the Tupi people, a group indigenous to Brazil. In the Tupi language, "jaborandi" means "what causes slobbering." The natives used the plant to fight fire with fire, to induce sweating in the treatment of fever. An infusion of the leaves was also used for liver and skin conditions and gastrointestinal and respiratory ailments. When jaborandi crossed the boundary into Western medicine in the 1870s, pilocarpine emerged as a novel cure-all, the ideal medicine for illnesses ranging from acute rheumatism to tetanus, fever, and asthma. Many of these applications later fell into disuse, with their ineffectiveness ultimately being exposed. The use of pilocarpine for glaucoma, however, persisted, having generated substantial evidence for its value in the condition's treatment.

But somewhere along the way, it became assumed that jaborandi would reliably produce pilocarpine eternally for medicine. Other

than some sparse attempts at generating the compound synthetically in the 1970s, the history of pilocarpine in the twentieth century is virtually silent. That is, until the 1990s, when, after decades
of harvesting of *P. microphyllus* in the wild, the plant found itself
uncomfortably close to the precipice of extinction. Human harvesting practices have affected wild populations of *Pilocarpus* in two
ways. First, the zealous plucking of leaves, which are dried and
then sent to pharmaceutical companies for pilocarpine extraction,
has reduced wild plant populations. Without their leaves, the plants
can no longer capture sufficient sunlight to produce energy for survival. Second, human harvesting has influenced recent environmental adaptations in *Pilocarpus*, with some species growing to smaller
sizes and having fewer leaves than their ancestor plants. Reduced
growth means reduced synthesis of pilocarpine, which in turn means
smaller quantities of the compound for pharmaceutical production.

One of the hurdles that must be overcome in finding ways to
bump up pilocarpine production through synthetic methods involves
first understanding how the substance is synthesized naturally in jaborandi. Despite more than a century of the compound's use in conventional medicine, knowledge of this process is really only in its infancy.
Once more becomes known about its natural synthesis, pilocarpine's
generation on a commercial scale using a synthetic or semi-synthetic
lab-based approach may become possible. In the meantime, in Brazil, where the future survival of *P. microphyllus* plants in the wild is
under considerable threat, plantations for the sustainable cultivation
of jaborandi are being developed. Out of all this work, however,
a complex situation has emerged. A synthetic product that would
relieve the threat to overharvesting could also result in lowered
demand for cultivated plants, thereby jeopardizing the livelihoods
of Brazilian *Pilocarpus* farmers. Avoiding this potential problem will
likely involve diversification on both sides, with farmers relying on
varied crops for income and with some pharmaceutical companies
employing synthetic approaches and others continuing to rely on
pilocarpine extraction from leaves. Such a situation, however, is
entirely hypothetical, since lab-based production of pilocarpine on
an industrial level has not yet been realized. Furthermore, although
plantations are being created, the future of wild *P. microphyllus* and
its relatives remains precarious. Local peoples who have utilized

Pilocarpus for various purposes continue to rely on the plants to fulfill their own needs.

Some of the world's most effective anticancer drugs have come from plants. In the Indian Ocean southwest of Mozambique, Africa, lies the island of Madagascar, the natural home of Madagascar periwinkle. But long ago, *Catharanthus roseus* escaped from its native land. It now thrives in the wild in many regions of the world and is a popular perennial garden species, known for its pinkish-purple flowers and dark green, oval-shaped leaves. In 1958 and 1959, chemists at the University of Western Ontario in Canada and at Eli Lilly and Company in the United States independently discovered a compound in the periwinkle that inhibited the growth of cancer cells. Known as vinblastine, this compound represented the first of the so-called vinca alkaloids. The second, vincristine, was discovered shortly thereafter. Both compounds undermine the reproduction of cancer cells by binding to tubulins, proteins that normally polymerize during cell division to create a structure needed for chromosome separation. Vinca binding prevents the splitting of chromosome pairs for replication and eventual partitioning into new daughter cells. The cancer cell, permanently stuck in the replication process, ultimately dies.

The subtle anomaly in chemical structure between the two vinca compounds produces important differences in their activity and toxicity. Vinblastine is used to treat Hodgkin lymphoma, a cancer of the lymphatic system that usually subsides following therapy. It is also sometimes deployed to do battle against testicular and breast cancers and Kaposi sarcoma, a cancer of connective tissue. Vincristine is used to treat breast and lung cancers, leukemia (a cancer of blood-forming tissues), nephroblastoma (a childhood cancer), and Hodgkin and non-Hodgkin lymphomas. When it comes to toxicities associated with the vinca alkaloids, their profiles diverge substantially. Whereas vinblastine can cause drastic reductions in white blood cells, leaving patients with little immune defense against infectious organisms, vincristine has less of a suppressive effect on white cells but can cause severe abnormalities in the function of nerve cells, which may manifest as pain or tingling in the extremities or loss of hearing or vision.

Madagascar periwinkle was used medicinally long before the discovery of the vincas. On its native island, it was used by locals to arrest bleeding and to treat symptoms of diabetes-like conditions. It also appears to have been used as a sedative. Following dissemination of the plant to other parts of the world, it found applications in the treatment of asthma, tuberculosis, malaria, sore throat, and high blood pressure. In the Philippines, an abortifacient concoction was made from the plant's roots. In Jamaica, its leaves were steeped to produce periwinkle tea, claimed to relieve the symptoms of diabetes. In 1952 Canadian scientist Clark Noble learned of this remedy and sent word to his brother, Robert, at the University of Western Ontario. Curious about the compound in Madagascar periwinkle that might account for the tea's antidiabetic activity, the latter Noble set to work investigating the plant and isolating compounds from it. In 1958, when he and Charles C. Beer reported the discovery of the first vinca alkaloid, they were searching for a compound that would be effective against diabetes, not cancer. And despite decades of research, the efficacy of the vincas in the treatment of diabetes has remained unsubstantiated. In contrast, research into their ability to fight cancer continues to prove fruitful to this day. Synthetic methods for the production of vinblastine and vincristine have been developed. Particularly for vincristine, however, the synthetic versions are generally less effective than the naturally occurring product.

The ability of plants to relieve pain has had, for many centuries, a remarkable effect on human lives. Among medicine's most ancient and best-known pain relievers are plants like willow and opium. The earliest written record of the medicinal use of plants now known to produce the painkiller salicylic acid dates to around 3000 BCE, in the form of inscriptions describing remedies made from willow trees in a Sumerian stone tablet recovered from the lands of modern Iraq. An extract prepared from the leaves of myrtle that was used to relieve pain in the abdomen and back in pregnant women was mentioned about 1,500 years later in the *Ebers Papyrus*. Renowned Greek physician Hippocrates suggested concoctions from the bark and leaves of willow trees for the mitigation of fever and pain associated with childbirth, and similar treatments were recommended by Greek and Roman physicians in later centuries. Elsewhere in the world the bark and leaves of species of willow, poplar, and myrtle, and certain parts of spirea plants, were employed in various tonics and decoctions for controlling inflammation and pain.

Aspirin's modern history began in 1828, when salicin, from which salicylic acid is derived, was isolated in small quantity from the bark of *Salix alba* by German chemist Johann Buchner. The following year, French chemist Henri Leroux invented a method to extract salicin in crystalline form from willow. But it was the breakthrough by Italian scientist Raffaele Piria in 1838, in finding a way to chemically convert salicin to the more potent salicylic acid, that aided the subsequent development of methods to generate the acid synthetically. The first do to so was French chemist Charles Gerhardt, who in 1853 produced acetylsalicylic acid (ASA), which remains the chemical of modern synthetic aspirin. Gerhardt's early formulation of ASA, however, drew little attention, and some forty-five years later, chemist Felix Hofmann, working for Friedrich Bayer in Germany, reinvented Gerhardt's compound. Hofmann did so by attaching an acetyl group to a form of salicylic acid that had been generated based on a method developed by fellow countryman Hermann Kolbe in 1860. Kolbe's compound caused severe stomach irritation, but Hofmann's acetyl group did not possess such intolerable side effects, and thus in Hofmann's invention, Bayer aspirin was born.

Aspirin is a paragon in the pharmaceutical industry. It was the first synthetic pain reliever developed, and constant research into its mechanisms of action and biosynthesis in nature has led to its diversification into therapy regimens for cardiovascular disease and cancer. Aspirin is a nonsteroidal anti-inflammatory drug, or NSAID, as distinguished from the steroid anti-inflammatory drugs known as glucocorticoids. NSAIDs relieve pain and reduce fever and inflammation. These effects arise from a single cellular action—the inhibition of cyclooxygenase (COX), an enzyme that comes in two main varieties, COX-1 and COX-2. The acetyl form of salicylic acid is in many ways an artful disguise, since although aspirin enters the body in the chemical form of ASA to ease its effects on the stomach, after its absorption through the wall of the gastrointestinal tract it is metabolized by enzymes in the blood plasma and liver to salicyclic acid. It is this form that is responsible for aspirin's physiological activity, which is realized through its ability to put the brakes on COX activity.

Under normal circumstances (in the absence of aspirin), COX mediates the production of substances called prostanoids. These molecules augment the inflammatory response of tissues by sending out cellular signals that promote swelling, pain, and fever. When salicyclic acid is present, however, the transmission of these signals ceases, which explains why aspirin is so effective in controlling the hallmark symptoms—pain and inflammation—associated with conditions such as arthritis. The popularization of low-dose aspirin for preventing blood clots is attributed to the disruption of the prostanoid pathways. In recent years, aspirin has also drawn attention for its potential value in the prevention of breast cancer in women at high risk. This ability, too, stems from its inhibitory effect on COX. A prostanoid dubbed prostaglandin E2 is capable of increasing estrogen production within breast tissue, and this appears to be a factor in the development of estrogen-dependent breast cancer. Following administration of low doses of aspirin, however, estrogen levels in the breast are reduced, and the risk for this type of cancer, which accounts for some three-quarters of all breast cancer cases, may be cut by as much as 20 percent in women taking the drug.

Not all NSAIDs are created equally, and certainly aspirin, due to its effectiveness and its relative safety, is the favorite of the bunch. Scientists' appreciation for salicylic acid has expanded dramatically in recent years, not only because of its newly discovered applications in medicine but also because of research that has turned up new information about its biosynthesis and biological activity in plants. Salicylic acid is a secondary metabolite. It has been detected in leaves and other plant parts in species ranging from crabgrass to barley, rice, and soybeans, in addition to the medicinally relevant spirea and willow. The compound is involved in the induction of flowering in certain plants, and it possesses allelopathic activity when released into the rhizosphere of a plant's roots (the rhizosphere is the region of soil encompassing the roots). Allelopathy in this instance appears to result from the compound's ability to inhibit the uptake of specific nutrients by competitor plants. The acid appears to defend plants in other ways too. For example, following infection with a pathogen, some plants increase their biosynthesis of salicylic acid, apparently in an attempt to ward off the offender.

Salicylic acid also has the truly unique ability of thermogenesis, which is found only in certain members of the Araceae family, a group of plants distinguished by the presence of a flowering structure known as a spadix (the equivalent of an inflorescence, or flower cluster, in other types of plants). The thermogenic effect was first observed in the voodoo lily. In this species, the acid generates heat during anthesis, the time when the plant's flower is open and actively attracting insects for pollination. During anthesis, *Sauromatum guttatum* gives off an odor that is offensive to animals (the smell is often described as putrid meat) but that allures a variety of pollinators. The production of the volatile odor-generating compounds coincides with a rise in temperature in a structure within the plant's spadix known as the appendix. The temperature of the appendix can increase by as much as 57.2°F over air temperature during the daytime stretch of anthesis, only to cool off again as night approaches. Salicylic acid production increases dramatically in the voodoo lily appendix in the day before anthesis, and the plant also appears to become more sensitive to the compound as flowering nears. The appendix is heated up by the acid presumably through the chemical's stimulation of an alternative energy pathway within the cells of the structure.

The wealth of knowledge of the natural, biological role of salicylic acid that has poured forth in recent years has put a new twist in the intrigue that this substance has generated since its discovery. This abundance of information stands in stark contrast to pilocarpine, a compound whose chemistry and biological functions were virtually ignored in the realm of research while its medicinal relevance persisted and increased throughout the twentieth century. The status of the natural sources of these two compounds reflects their disparate histories in medicinal chemistry. The diligent efforts of nineteenth-century chemists to develop a synthetic process for ASA production have proven invaluable in circumventing the potential loss of plants through overharvesting. Although salicylic acid is found in many more species than is pilocarpine, the worldwide distribution and consumption of aspirin are staggering—far greater than pilocarpine could ever experience. An estimated 80 million tablets are consumed each day in the United States alone. At that level of demand, had the invention of synthetic aspirin never occurred, the world's willows, poplars, myrtle trees, and spirea shrubs, if not already extinct, would find themselves in the same situation as *P. microphyllus* in South America: at the outpost of existence.

More potent than aspirin and powerfully addictive is morphine, which is found in extracts of opium poppy and has come to fill a crucial yet controversial role in medicine. Opium poppy is native to Anatolia (Asia Minor), the modern-day Asian continental region of Turkey, but is now found in many parts of the world, including India, Southeast Asia, Europe, Africa, and the Americas. While *Papaver somniferum* grows in the wild in many of these places, it is also cultivated. It is the source of raw opium, about 10 percent of which is Heroin No. 1, or crude morphine. Similar to aspirin, the history of opium appears to begin with the Sumerians, who called the plant *hul gil*, meaning "joy plant." This seems a clear indication that the Sumerians were aware of the plant's feel-good effects. Opium poppy was mentioned in the *Ebers Papyrus*, though only as a way to calm crying infants, and appeared later in ancient Greek medicine, where it was used as a hypnotic sleeping potion. The name "opium" comes from the Greek *opos*, describing sap or sticky juice. The processes of slicing opium seed capsules to drain the sap and of drying the sap to produce opium powder were detailed by Greek physician Dioscorides in his *De Materia Medica* in the first century CE; his methods are still used today. Dioscorides, along with others, including Pliny the Elder, knew of the inevitable, overpowering tranquilizing effects of opium following high doses. Other Greek physicians, including Galen of Pergamum, prescribed the drug to reduce pain but, aware of its risks, discouraged its liberal use. In the lands of the ancient Middle East, opium became widely used, having taken the place of wine in the Muslim world.

In China, opium was employed as a sedative by surgeon Hua Tuo in the second or third century CE. About 1,400 years later, the practice of smoking the substance proliferated within the country, and following the upsurge of opium imports into China by the British East India Company in the eighteenth century, opium earned a reputation for abuse. The company's activities were fundamental in perpetuating China's addiction to the drug, and disenchanted with the destruction that this addiction was having on its citizens and economy, the Chinese government attempted to stamp out the company's rampant, illegal operations. This in turn fueled the infamous first (1839–1842) and second (1856–1860) Opium Wars in China.

In 1804, prior to the wars, German chemist Friedrich Wilhelm Adam Sertürner isolated what he believed was the compound responsible for the sleep-inducing effects of opium poppy extract. Sertürner repeated the isolation in 1817 and named the crystalline substance morphine. Its chemical structure, however, was not elucidated until the early 1900s, after which several semi-synthetic and synthetic derivatives were developed, including buprenorphine and hydromorphone, as well as the illegal substance heroin. Morphine came to serve as a synthetic source of codeine, another powerful pain reliever.

Similar to quinine and pilocarpine, morphine is an alkaloid, the principal of some forty different alkaloids that have been identified in opium. Morphine produces its pain-relieving effects by triggering the activity of opioid receptors in the central nervous system. There are three known subtypes of opioid receptor, and the one to which morphine is most attracted—the mu receptor—mediates not only the drug's pain-relieving capabilities but also its induction of a euphoric state, depression of respiration, and physical dependency. As a result, morphine's pain alleviation is inseparable from its addictive and psychological activities. In addition, morphine tolerance necessitates increasing doses to achieve pain relief. Certain morphine derivatives, such as buprenorphine, effect potent pain relief but have a reduced tendency to tolerance and addiction. The disparity between these derivatives and morphine is caused by differential effects on opioid receptors. Morphine, for example, is an agonist at mu receptors, meaning that it prompts the receptors to generate the same cellular response that they would in the presence of their natural endogenous binding partner (an endogenous substance is produced naturally within the body, in contrast to a drug, which is exogenous, from outside the body). Buprenorphine, however, is described as a partial agonist at mu receptors because, though it is attracted to the receptors, it only weakly stimulates them, producing a muted response. In addition to its use as a pain reliever, in 2002 buprenorphine was approved by the FDA for use as an opiate withdrawal therapy. Its reduced triggering of the opioid response enables addicts to step down gradually from abuse of opiates such as heroin, lessening the severity of withdrawal effects.

The chemical constituents of opium are among the most intensely studied compounds in pharmacology, and as a result, there is a vast amount of literature available on their physiological effects. In the twentieth century, opium attracted scientific attention because very little was understood about how the drug acted on the body and why it could produce the simultaneous euphoric, pain-relieving, and addictive effects. Most of this centered around the need to identify the receptors that the drug interacted with on cells. Receptors occur in parallel with endogenous binding partners, and so opioid receptors evolved in humans because the human body produces an opium-like substance. The first opioid receptors were discovered in the early 1970s and shortly thereafter the first endogenous opioid molecules were reported. These molecules are responsible for the phenomenon of the "runner's high," substances known scientifically as endorphins and enkephalins. Both occur in the brain and central nervous system, and while endorphins produce a "feel-good" effect, enkephalins are potent pain suppressors. Endogenous opioids were later found to influence other physiological responses, and their activities are now understood to be quite varied and complex.

Similar to salicylic acid and pilocarpine, the biological role of morphine in plants is much less well characterized than its role as a drug. In opium poppy, the sap, or latex, extracted from the seed capsules represents the cytoplasmic contents of laticifers, secretory cells that act as latex-conducting pathways. Laticifers and adjacent structures known as phloem sieve elements, which conduct the flow of phloem from cell to cell through a specialized sieve tube, have been associated with the biosynthesis of poppy alkaloids, as well as with their accumulation in the capsule latex. The actual plant bioactivity of morphine, however, remains uncertain. Morphine may come to the aid of the opium poppy following mechanical stress on its seed capsules. Such stress triggers morphine's metabolism into bismorphine, which binds to chemical residues in pectins (structural components in the walls of plant cells). Bismorphine binding establishes vital connections between pectin molecules, linking them together and strengthening the wall. The accretion of bismorphine within the injured capsule also blocks the activity of pectinase, an enzyme that normally dismantles the chemical integrity of the wall. Because mechanical stress is usually one of the first events in pathogen infection, the morphine to bismorphine response may be a first-line defense against capsule invasion by disease-causing agents.

Because morphine can be produced through total synthesis for legitimate medical purposes, the opium poppy is not threatened by modern pharmaceutical demands. In fact, cultivation of the plant for the production of illicit raw opium has served as a motivating factor in discouraging the planting of opium poppy, even as a garden species. Although morphine has acquired a shady reputation due to its inclusion in the ever-growing list of drugs of abuse, it has a strong toe hold in conventional medicine. In what is perhaps its most noble role, it serves as a last resort, welcomed with great reluctance, in the relief of suffering for cancer patients whose conditions have advanced beyond the reach of modern medicine.

Bioprospecting and species discovery are fundamentally ancient practices that in modern times have evolved alongside technological developments that reveal organisms' most basic components, their genes, proteins, and other chemical constituents. These technologies also circumvent the need to harvest copious specimens in the wild, which has proven valuable in helping researchers gain permission from regional or national governing bodies to perform natural-products discovery within the boundaries of protected and managed lands. Developments such as the National Parks Omnibus Management Act of 1998, which authorized benefits-sharing agreements between the US National Park Service and scientific institutions or private companies, have further facilitated discovery. Under the Omnibus Act, both national parks and research institutions profit from the discovery of commercially valuable products found within park boundaries.

Substances found in nature are so extraordinarily valuable because they form the foundation of drug development. The number of compounds, raw or synthetic, that make the final cut for commercial drug development, however, is surprisingly few. Thousands upon thousands of natural, semi-synthetic, and synthetic substances have been evaluated using routine high-throughput screening techniques. But only a small percentage has made it into medical practice. In the case of natural products, some are not medically useful because they cause harmful side effects in humans. In many cases, however, natural substances simply lack potency. Using synthetic chemistry, slight tweaks can be made in a compound's structure, eliminating the chemical constituent responsible for side effects or introducing a side group that increases potency. But while synthetic chemistry has worked wonders in this respect, a heavy dependency on the generation of new compounds using synthetic approaches, namely combinatorial chemistry, has spawned far more useless compounds than effective ones. Furthermore, this process still depends on an initial parent compound, often a naturally occurring substance.

Much of current natural-products discovery centers on the identification of species that have been used in traditional medicine for centuries, so ensuring a continual supply of novel products to feed into synthetic chemistry is the work of not only bioprospectors—individuals who search for substances in nature—but also ethnobotanists and ethnopharmacologists. These experts investigate various aspects of plant science and human interactions with plants. Their efforts are further supported by another element key to the success of modern drug development, the discovery of new plant species, work performed in large part by modern-day plant hunters.

Back in the days of Banks and Hooker, plant-hunting expeditions were concerned with the discovery of new species, particularly exotics, and with the collection of seeds. Naturalists needed to find funding, hire transport, and work with individuals and sometimes political officials in foreign lands. Today, the endeavor remains in these respects very much the same, involving not just one naturalist or one researcher from one university, but multiple collaborating scientists, with expertise in different areas, and the cooperation of local peoples and national institutions that determine the areas where field research is permitted. And, similar to the plant hunters of centuries past, those committed to finding new species in the modern era share a deep respect and appreciation for nature, compelling them to spend countless hours submerged in rain forests or exposed to the dry heat of deserts or the frigid temperatures and harsh winds high up on mountain slopes.

Modern plant hunters have been very busy. An estimated 2,000 new species of plants are discovered globally each year. Since its inception in 1759, the Royal Botanic Gardens, Kew, has been one of the most important leaders in the process of finding and identifying new plants. Many of the big names in the history of plant hunting—Banks, Hooker, Francis Masson, George Forrest, Scottish missionary and explorer David Livingstone—gathered hundreds of specimens for Kew, a number of which still exist in the Gardens' collections. Modern-day Kew scientists are just as determined as ever to seek out new species. In 2009, as part of the organization's Breathing Planet Programme, scientists and volunteers working in 100 countries in association with the Gardens discovered 250 new species of plants and fungi, 62 of which were found in the rain forests of Borneo, part of the Sundaland biodiversity hotspot. Many of the Borneo species new to science included various types of flowering plants in the custard apple family, different kinds of orchids, and several new species of beautyberry. More than 30 previously unknown species were identified in Madagascar, including new wild coffee plants that produce extraordinarily large beans. Other discoveries included two new species of eucalyptus in Southwest Australia, a new species of yam in South Africa, and new species of lily in Iran and myrtle in South America. A new member of *Strobilanthes* in India and another in the Lesser Sunda Islands, southeast of Indonesia, are likely to contain important natural products for medicine. Several well-known species of *Strobilanthes* are used in traditional systems of medicine, and bioactive compounds, including substances with antioxidant and antidiabetic activities, have been isolated from them and investigated clinically.

The discovery of new plant species is an enterprise that beckons attention on an international level, and certainly the collective global effort amounts to the identification of many new species each year. Researchers affiliated with academic or government institutions fill an important role in this process. But the identification of so many new species stems in large part from the interest of small, local organizations, particularly local nurseries, gardens, and conservation groups, as well as individual volunteers within communities. Much of Kew's success has come as a result of interaction with citizens and researchers in local communities who understand and appreciate the need to identify the many forms of life hidden in Earth's ecosystems.

There are many examples of small organizations that fill a vital role in plant discovery and conservation on local levels. The work of many of these organizations has translated into progress in plant conservation internationally. In the United States, individual-run groups like Peckerwood Garden Conservation Foundation in Texas and Quarryhill Botanical Garden in California have made immense contributions to plant science. In the case of Peckerwood and Quarryhill, each maintains an elaborate collection of plants at its local facilities, with the former focusing on native species in southwest Texas and across the border into Mexico and the latter focusing on Asian plants. Horticulturalists at both gardens have made multiple expeditions in search of new specimens, and their efforts to identify new species, to collect and share seeds, and to propagate endangered plants in protected gardens have been fundamental to plant conservation. Peckerwood has worked to raise awareness of endangered plants in both Mexico and the United States, cultivating an appreciation for native species through expeditions to various parts of the region and through educational programs. In emphasizing the artistic and environmental elements of horticulture, Peckerwood strives to bring greater local attention to the impacts on plants of activities such as overgrazing and development. Quarryhill shares a similar philosophy and has engaged in more than twenty expeditions to China and multiple expeditions to Japan. Quarryhill also shares discoveries and materials with other plant scientists, knowing that their contributions to horticulture are fundamental to conservation and to expansion of the basic scientific knowledge of plants.

By bringing together two very different, yet interrelated elements of plants—their ecological significance and their value to medicine—the extensive knowledge gained from botanical and pharmacological research has strengthened the cause for nature conservation. The search for new species and new natural products transcends all ecosystems, from rain forests to deserts to oceans, and has emphasized the need to protect these areas. In addition, research into the biological functions of plant compounds has improved our understanding of how plants interact with other organisms. Beyond this, natural-products discovery also has the potential to bring new meaning to medicine. As researchers dig deeper into the histories of the people and the plants behind the many ancient systems of healing, the interests of conventional and traditional medicine are likely to converge. The harmonization of these two approaches stands to significantly advance human health on a global scale.

But there are a number of issues facing the discovery and commercialization of plant-based medicines, particularly for agents whose development stems from the practices and knowledge of indigenous peoples. The discovery of compounds that could lead to medicines for currently untreatable diseases and that could benefit people worldwide is a noble venture. But for this process to truly benefit all who are invested, whether through scientific resources or through an ancient connection to the plants and land, there must exist recognition and respect for the art of ancient healing as a form of medicine and for the scientific exploration of natural products as a source of agents that billions of people globally could come to depend on. As a result, we must learn to coexist on two levels—with one another and with all Earth's other forms of life.

7

Learning to Coexist

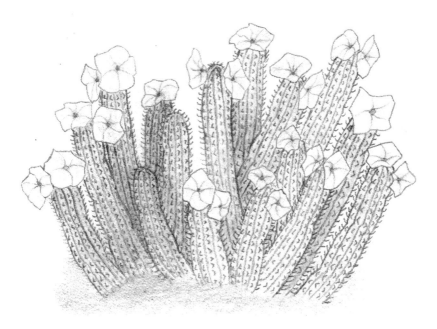

Hoodia gordonii, a succulent native to southern Africa.

THE DISCOVERY AND DEVELOPMENT OF new plant-based drugs are imperative to the advance of medicine and the improvement of human health. But the process of identifying and characterizing new plants and plant compounds, and the subsequent development of compounds into commercial entities, is an ethically and legally

143

conflicted area of research, mainly because it treads on the territory of traditional knowledge.

Since the mid-1960s, various laws have been enacted to protect the folklore, arts and crafts, and traditional knowledge of indigenous peoples from exploitation. A precedent was set by the African Intellectual Property Organization (OAPI) agreement of 1977, which declared traditional medicine and knowledge passed from generation to generation within any African ethnic community to be property of the community. Ever since, traditional knowledge and intellectual property laws have repeatedly found themselves in conflict.

Controversy over bioprospecting and the development of drugs is intimately associated with intellectual property laws. Between countries, laws dictating what can and cannot be patented vary. Inventions are broadly considered a form of intellectual property, and as such they are generally granted protection under patent law. The intention of patent laws is to provide incentives for inventors, encouraging new inventions, perpetuating the invention cycle. Patents provide a mechanism whereby inventors disclose information about their work, making known to all the existence of the invention and the details for how it can be copied. In return for disclosure, inventors receive patent protection for a designated period of time, during which they are free to profit from their inventions. But the process isn't perfect, and in some areas of science, patents may actually hinder invention, rather than act as a motivating force. The rippling effects of this are felt to an exponentially greater degree in developing countries, where the extraction of local natural products and knowledge by patents awarded to researchers from wealthy countries has crippled cultural heritage and creativity.

Attempts to secure patents on natural entities originated with Louis Pasteur in 1873, when the US Patent Office awarded Pasteur ownership over a strain of yeast employed in beer making. In 1937 Pasquale Joseph Federico, head of the US Patent Office, stated that Pasteur's award likely would not have been granted by the office during his tenure, "since it may be doubted that the subject-matter is capable of being patented." This, however, came only shortly after the Plant Patent Act of 1930, which allowed asexual, cultivated plants, with the exception of tubers, to fall under the protection of patents. But Frederico was not alone in his aversion to the notion of someone bearing ownership over nature. Up until the Plant Patent Act, it seemed to have been assumed generally that plants were not patentable things. Even during all the years that passed between the Patent Act of 1793 and the Plant Patent Act of 1930, there was only one application submitted for a patent on plant material. Seeking protection for a fiber from the needles of a pine tree, the application was denied.

But the ability of scientists to invent new types of organisms from existing ones has presented significant challenges to patent law. Patents sought on living organisms became an increasingly contentious issue throughout the last century. In 1970 the floodgates broke in favor of plant patents, when sexual plant cultivars became eligible for protection, a measure followed several years later by the first patent on plant seed. These developments, however, were overshadowed in 1980, when the ruling given in the Supreme Court case *Diamond v. Chakrabarty* opened the door to patents on human-made organisms. This resolution arguably altered the course of science. The ability to engineer an already existing life-form into a profit-making asset sparked an upsurge of activity in the biotechnology industry. The *Chakrabarty* case now is perhaps more infamous than famous. Ever since the ruling, patents have been filed for everything from genes to genetically modified organisms. Protection for new types of plants invented through breeding techniques is now guided by a distinct set of international intellectual property laws.

Under US statutes, invention *or discovery* is patentable, as long as it is "nonobvious" and has utility. Genes, natural products, and other natural substances that have been isolated and purified are candidates for patent protection only if they have a legitimate use. Without this, patent rights will not be granted. The utility element is a controversial part of patents. Had the substance for which a patent is sought not been isolated and purified, certain uses for which patent protection is filed may not have been discovered otherwise. Such substances are, under US patent law, considered unique from their naturally occurring counterparts. Whether a gene or a plant compound, it is essentially out of its natural element and therefore presumably does not function "exactly" as it does in nature. The irony of this is that, later, in abating public concerns over the safety of a product, companies often claim the opposite—that their new biopesticides and genetically modified plants are not at all different from their natural counterparts.

Genes and plant compounds in nature do not infringe on their patented analogues. But many individuals and companies seek extremely broad rights to their products. A genetically engineered crop that finds its way to a neighboring farmer's property, taking root there under its own power or mixing with nonengineered plants, suddenly places the farmer in violation of patent rights. On an international level, and particularly for patents on plant-based medicines, infringement often comes as a surprise—many indigenous peoples and governments, for example, do not know that patents exist on natural products sourced from species that they have long relied on for medicine or food.

The utility clause has raised serious ethical concerns in natural-products discovery. Plants and the medicinal substances they contain are part of human cultural heritage. While researchers might venture into nature and select plants at random to investigate for natural-products discovery, the more effective approach is to study plants that have been used in traditional systems of medicine. This increases the chances of isolating a compound with potential for drug development. The problem with this is not necessarily about the minutiae of claiming ownership over an isolated plant substance. Rather, the argument encompasses the far larger problem of enabling intellectual property claims to engulf traditional knowledge.

The commercialization of natural products and traditional knowledge is moderated primarily by the World Trade Organization program known as Trade-Related Aspects of Intellectual Property Rights (TRIPS). Ownership issues that impact biodiversity overlap with the domain of the Convention on Biological Diversity (CBD). The CBD ensures the conservation of biodiversity and the sustainable use of biological resources and recognizes indigenous ownership of folklore and traditional knowledge. It also promotes the equitable sharing of knowledge and genetic resources with researchers and industries. Equitable sharing is accomplished through cooperative efforts for developing and applying commercial technologies. This is the same process used by the US National Park Service, in which benefits-sharing agreements ensure royalties for all parties involved, in the event that a product achieves commercial success. But whereas the CBD extends ownership of traditional knowledge to cultural groups, TRIPS recognizes ownership only under patent, and anything that is not patented is by definition open to exploitation.

In recent years, TRIPS has worked to address the concerns of developing countries lacking the production capabilities necessary for development and commercialization of a natural substance. The program has built in compulsory licensing options for these countries, such that the patent holder must share production rights with the country or state. This allows a generic version of a drug to be used domestically, with the patent holder receiving payment for the copied version. TRIPS has also attempted to bring its perspectives into closer association with the CBD, working with member countries to more clearly define what constitutes traditional knowledge and to improve arrangements for benefits sharing and material transfer. Other proposed improvements include requiring patent proposals to disclose the source of knowledge or biological material from which a product has been derived and to demonstrate prior informed consent, meaning that the country or state in which the product was discovered was aware of the intent of the research conducted there.

The lack of prior informed consent forms the basis of biopiracy claims. In the past couple of decades, watchdog organizations such as GRAIN (Genetic Resources Action International), a nonprofit group that supports community-based control of local natural resources, have defended small communities from the social harms of bioprospecting, the most offensive of which is biopiracy. "Stalking" and "plundering" of natural resources by bioprospectors, particularly in instances where researchers seek out new natural products in developing countries, are terms frequently used to describe biopiracy, which also includes the swiping of traditional knowledge from indigenous peoples with intent of commercialization. Because of this, the reputation of bioprospecting has been marred, causing many to overlook its potential. The discovery of drug compounds in plants has the power not only to advance medicine but also to advance the conservation of biodiversity and foster economic development within countries that have a rich supply of natural products but have no means of developing them.

Bioprospecting requires close regulation and monitoring. Though laws are in place for binding researchers and industries to good bioprospecting practices, enforcing regulation has been difficult. The exploitation of natural resources and peoples has been practiced for centuries, and historically many viewed these practices as acceptable, even normal. But changes in modern societal values have confused the traditional norms of exploitation. Indeed, many people hold an ambivalent viewpoint on whether such activities are ethically and morally right or wrong.

Global capitalistic practices thrive on competition for invention and discovery. The ever-hastening pursuit of technology and progress has left a wide wake lapping on the shores of human heritage, gradually eroding away the cultural diversity of indigenous peoples and replacing traditions with often materialistic things. In multiple instances since the late 1990s, indigenous groups have fought for retribution against patents awarded on biopiracy of natural products and traditional knowledge. Substances pirated for commercial profit have included both over-the-counter herbal supplements and compounds under investigation for development into approved pharmaceuticals. Cases have included an appetite suppressant discovered in *Hoodia*, an aphrodisiac and fertility-enhancing substance isolated from maca, and various herbal medicines made from compounds found in neem trees. Indigenous peoples and nongovernmental organizations have sounded the alarm for multiple reasons but primarily because of the absence of informed consent, a problem that stems from flawed patent systems that consequently guide good scientists down ethically conflicted paths.

Prior informed consent and benefits sharing are not obligated by patent law. In voicing their concerns, indigenous peoples are asking for a fair patent process, one that encourages transparency rather than one that encourages secrecy and sometimes even dishonesty. In the 1990s, as part of a systematic investigation to discover drugs from South Africa's indigenous plants, researchers working for the country's Council for Scientific and Industrial Research (CSIR) isolated a compound named P57 from *Hoodia*. These succulent members of the milkweed family are an important element in the diets of the indigenous San people. The San used the stems of the plant to quench thirst and suppress appetite when on long excursions in the desert. Two species, *H. pilifera* and *H. gordonii*, possess multiple compounds of interest to drug development, but P57 was the first to be patented. CSIR later licensed the patent to UK–based Phyto-pharm and sold the patent rights to Pfizer. The San, having learned of P57 and feeling that their traditional knowledge of the plant was leveraged unfairly, sought legal and financial retribution. In 2003 the CSIR agreed to equitable benefits sharing with the San, in the event that *Hoodia* products were commercially successful. The San were scheduled to receive a percentage of each milestone payment made to the CSIR by Phytopharm (these payments are made upon completion of successive technical stages of product development). The CSIR also agreed to give a percentage of royalties from the commercialized drug to the San once the product became available. While the battle over patent rights and benefits sharing was being hashed out in the courts, however, unapproved, over-the-counter supplements containing *Hoodia* extracts were pushed onto the market by herbal-products companies. Many of these products, unfortunately, did not contain *Hoodia* extracts though they were advertised as such.

More complex than the P57 patent were the three patents on maca that were awarded to different companies. *Lepidium meyenii* is cultivated in the Puna highlands in the Andean mountains of Peru and was once an important trade commodity for the indigenous Quechua-speaking peoples inhabiting the mountains. Starting in the 1980s, the Quechua began cultivating the plant, and since then, maca has filled an important nutritional role, with aphrodisiac benefits that have baited pharmaceutical companies.

The Quechua are not strangers to pharmaceutical exploitation. First, there was the discovery of the antimalarial properties of quinine, which was isolated from cinchona bark extracts based on knowledge that the natives used the plant possibly for medicinal purposes. In 1986 US scientists were given patent rights to ayahuasca, a vine native to South America and valued by the Quechua for its hallucinogenic properties. Protests by Amazonian nongovernmental organizations led to a patent annulment in 1999. Because the characteristics of *Banisteriopsis caapi*'s leaves were different from all other known descriptions, however, the patent was later restored. When the maca patents came along, the Quechua and Amazonian support groups had reached the limits of their tolerance for biopiracy.

Maca patent problems began in 2000, when Texas-based Biotics Research Corporation patented a peculiar concoction called "maca and antler for augmenting testosterone levels." The following year, New Jersey-based Pure World Botanicals, Inc. took out a patent on an extract of maca with intention for pharmaceutical development, and the year after that, Pure World received a second patent, this time for a novel alcohol-based extraction process tailored to maca. All three patents were opposed on the basis that the discoveries and inventions they claimed were not new at all; they had been a part of the local knowledge, traditions, and resources for centuries. Biotics and Pure World were sucked in mainly by the temptation of capitalizing on the aphrodisiac properties of maca.

Maca is known by various names, including Peruvian ginseng and Peruvian Viagra, and it is renowned for its ability to supercharge energy reserves and treat fertility disorders. It may also protect against physiological effects of stress, potentially strengthening one's resilience for life at high altitudes. These benefits probably have something to do with the plant chemistry that allows maca to grow at altitudes of about 13,100 to more than 14,700 feet, enduring cold temperatures and harsh winds.

Recent expansion of maca cultivation has been fueled largely by millions of dollars invested on the part of pharmaceutical companies and government food-based interests. In 1994 only about 0.2 square mile (128 acres) of land was devoted to growing maca. By 2002, the plant covered about 7.7 square miles (almost 4,930 acres). Yet, local land managers and indigenous groups were cut out of the picture when it came to the maca patents. Their labor in expanding the plant's cultivation and their knowledge of the plant's medicinal and nutritional benefits were not recognized. In a 2002 report, the Action Group on Erosion, Technology and Concentration (ETC Group; formerly RAFI, Rural Advancement Fund International), which supports human rights and cultural and ecological advancement and diversity, claimed that according to current laws, patents were fundamentally incapable of acknowledging traditional knowledge and "informal innovations of indigenous people." The group pointed out that US patent law, the most widely employed system, conflicts with the efforts of the World Intellectual Property Organization, which is working to bring traditional knowledge into the patent equation.

The race to patent ideas and discoveries has cut off the flow of communication in scientific research. In biotechnology and pharmacology, scientists frequently refrain from providing much detail about their work for fear that someone might steal their ideas. Yet, profiting from the creative concepts that have transcended generations and are embedded within indigenous cultures is often accepted without hesitation.

The threat to intellectual diversity generated by patents has created what Indian scientist and social and environmental activist Vandana Shiva described as knowledge monoculture. Shiva has been influential in bringing attention to the harm inflicted on indigenous peoples by misguided commercial and scientific interests. She has also increased awareness of the need to protect intellectual diversity globally. In *Biopiracy* (1997), she describes scientific knowledge as a tree, with basic research and discovery in a diversity of fields forming the tree's roots. But these roots, which grow and support the tree's trunk and branches, are, as she put it, "being starved, even as they are being rapidly exploited and harvested for profits." Shiva was one of the first to draw attention to commercially targeted research and its impact on the loss of diversity and extinction of scientific knowledge. The prospect of financial returns in commercially supported research generally triggers a shift in priorities, with social needs losing out to potentially profitable investments. Because the driving force of investment draws such extensive interest from research groups, the acquisition of knowledge in otherwise nonprofitable lines of discovery will be forgotten. In Shiva's words, these forgotten areas could ultimately "become extinct."

Diversity of knowledge is important for reasons that cannot be recognized by profit margins. Scientific knowledge from many different fields enables researchers to respond to environmental disasters and epidemics, to discover new species, and to identify species at risk of decline. Shiva's message was both cautionary and urgent: Placing profits ahead of practical needs and discovery creates an atmosphere supporting the destruction of intellectual diversity.

The indigenous peoples of the Andean mountains and Amazon rain forest are constantly struggling against industries seeking to claim intellectual property rights on some cultural or traditional element of their societies. Though the Quechua failed in their attempt to undo the patent bestowed upon ayahuasca, it did not deter them from pursuing retribution for the three patents on maca. Their goal, similar to the revisions in patent framework under consideration by TRIPS, was to establish requirements for the disclosure of sources of genetic materials.

The disputes over maca have been folded into a much larger battle between the indigenous peoples of Peru and the herbal industry. Peruvians have also fought patents on at least six other plants, known locally as chanca piedra, camu camu, hercampuri, yacón, caigua, and sacha inchi. Patents that have been refused for plants that are part of traditional Peruvian knowledge include a 1997 application for camu camu and a 2004 application for another maca product. In 2007 Peru submitted a document titled "Combating Biopiracy: The Peruvian Experience" to the Intergovernmental Committee on Intellectual Property and Genetic Resources, Traditional Knowledge, and Folklore in Geneva. The report outlined the biopiracy issues challenging the natural resources and wealth of biodiversity characterizing the country's lands. The purpose of the country's report was not to bring down the patent system. Rather, as stated in its concluding section, "it is simply a matter of achieving the objective of promoting innovation whilst maintaining a degree of fairness and equity among the stakeholders involved in the system."

Peru and Amazonian nongovernmental organizations have been accused of blowing the whistle too often. But their accusations are largely in reaction to patent claims that threaten the country's economy, biodiversity, and indigenous culture and traditions. Peru is working to shore up the ethical gaps associated with bioprospecting, and such proactive interventions could in turn reduce the exploitation of natural substances, especially by herbal-supplement companies.

India has had its own share of biopiracy problems and as a result has become a vital component in raising global awareness of biopiracy. The country has been successful in revoking multiple patents infringing on traditional knowledge. In 1995 a patent on the wound-healing compound curcumin, an alkaloid in the spice turmeric, was awarded to researchers at the University of Mississippi. The Indian Council of Scientific and Industrial Research countered the patent on the grounds that turmeric has been used for treating wounds in the country's indigenous systems of medicine for centuries. Because the patent was not protecting a novel discovery, it was ultimately revoked by the US Patent Office. Despite the absence of patentable material, which has produced a void in financial investment in curcumin research, scientists in academia have continued to study the compound. It has been found to possess anticancer and antidepressant properties, and synthetic analogues have been developed.

Another plant in India that has come up against patent issues is the neem tree (*Azadirachta indica*), which has long been valued in India for its insecticidal and antimicrobial characteristics. The extracts are incorporated into lotions and toothpastes and are used as contraception and for diabetes, leprosy, and skin conditions. Neem contains a wide array of bioactive secondary metabolites, one of which, azadirachtin, is found in the kernels of neem seeds and has drawn attention for its insecticidal activity.

Patent wars over neem originated with an azadirachtin extraction process developed in the 1970s and ultimately patented in 1985. Three years later the patent was sold to US chemical and materials developer W. R. Grace & Co., which used the process to develop a commercial neem-based insecticide. The company later struck a deal with an Indian firm to cultivate and process neem to fulfill market demand. In 1994, when another patent was granted to W. R. Grace & Co., this time in conjunction with the US secretary of agriculture, India fought back. Under question was the company's discovery of the fungicidal activities of neem extracts and the "novel" use of the extracts against fungal organisms, which was granted patent protection by the European Patent Office (EPO). As nongovernmental organizations and Indian organizations explained, these effects of neem did not represent new knowledge at all—Indian farmers had been using neem oil to prevent fungal growth on grains and crops for decades. In 2005, after a long battle, in which Indian officials provided substantial evidence to support their argument, the EPO revoked the patent.

India's perspective on natural-products patents is similar to that of Peru and other developing countries: There simply needs to be greater transparency and accountability in the patent application and review processes. In 1999 Indian officials initiated a plan to develop what they called the Traditional Knowledge Digital Library (TKDL). The TKDL arose in response to myriad patents for claimed novel discoveries and inventions that capitalized on India's traditional knowledge. Within several years of proposing the TKDL, the project moved from a plan on paper to an online database. Information on traditional medicines that had been passed orally through generations and existed in multifarious volumes was now available in one place and in a comprehensible framework, with translations of knowledge into multiple languages. In 2009 India opened the TKDL to the EPO.

Efforts like the TKDL are strengthened by the work of researchers whose projects are based on respect for traditional knowledge and open communication with indigenous peoples, local governments, and land managers. Such researchers are growing in numbers worldwide and are making significant progress in breaking down the walls that have been built between bioprospecting and indigenous peoples. But in order for the full potential of bioprospecting to be translated into reality, the rules dictating the game of natural-products discovery must be harnessed with a system of ethical checks and balances. Many indigenous groups have long possessed a deep-rooted resistance to the influence of Western culture, and although over the last several decades some of these tribes have become more open to communication and have even accepted the idea of building meaningful relationships with bioprospectors, natural-products discovery ventures headed by wealthy companies in the West have deeply wounded developing native peoples and developing countries, breaking their trust and offending their cultural heritage and traditions.

US- and European-led bioprospecting endeavors allowed to run free through the tropical forests of developing countries, because they find loopholes in patent laws or take advantage of existing oversights in funding and proposal-review processes, ultimately spawn animosity in the form of nongovernmental organizations and indigenous tribal councils when the intentions of their research become known. Activist groups serving in defense of indigenous and local communities have lashed out, assailing the pharmaceutical industry and the ideology of the West. These groups have encouraged the building of walls, dispelling hope for the free exchange of ideas through prior informed consent and benefits sharing, which could feed into the economies of indigenous communities, increasing their autonomy by enabling them to afford to pursue activities in conservation, sustainable farming, or the development of technology.

Anti-bioprospecting pressure from the inside, by national and tribal groups, has been especially problematic, leading to confusion on local levels and causing indigenous tribes who supported research to suddenly close the door.

A project begun in 1998 known as "Drug Discovery and Biodiversity Among the Maya of Mexico," a collaborative effort headed by University of Georgia scientist Brent Berlin and supported by a $2.5 million International Cooperative Biodiversity Groups (ICBG) grant awarded by the Fogarty International Center on behalf of the US National Institutes of Health (NIH) and the National Science Foundation (NSF), illustrates the issues that ethnobotanists and ethnopharmacologists face due to the stigma of natural-products discovery and the pharmaceutical industry. The goals of the project were to identify native Mexican plant species with medicinal potential and to then characterize bioactive substances found in the plants.

Mexico is home to a broad diversity of plant and animal life, with the greatest variety found in the country's central and southern regions. Urban sprawl and various human activities that have disrupted local ecosystems, however, have placed native plants at high risk of loss in the coming decades. Berlin and his team of University of Georgia scientists were operating fundamentally on the notion that their work, bioprospecting in Mexico's most biodiverse lands, would benefit both medicine and conservation, finding new drugs to treat human diseases and saving species before it was too late.

Ethnopharmacologists planned to focus their research on the Maya and their local plants. The Chiapas Highlands, or Sierra Madre de Chiapas, located in Mexico's southernmost state, were identified as an important area for natural-products discovery. Though perhaps best known for its impressive collection of endemic amphibians, particularly salamanders and newts, the mountainous slopes and broad plateaus are also home to numerous plant species. Researchers planned to screen the isolated compounds for bioactivity in a newly developed lab at Colegio de la Frontera Sur (ECOSUR) in San Cristobal. This would allow them to identify novel substances for potential development into anticancer drugs or agents capable of treating diabetes, cardiovascular diseases, and even HIV/AIDS. There were also plans to investigate methods of sustainable harvest for medicinal plants, crops, and species of interest to local horticulture.

To strengthen interactions with the local community, project coordinators planned to enlist the help of the Maya, training them in sustainable practices and research. Work opportunities for students from Mexico and the United States were also made available.

Researchers from the University of Georgia, in collaboration with the Chiapas-based ECOSUR and the UK–based Molecular Nature, Ltd., had received permits to conduct initial research in multiple municipalities in Chiapas and had applied for a bioprospecting permit in Mexico, the first permit to be allowed in the country under then newly established regulations. The researchers also set up medicinal-plant exploration partnerships with more than forty autonomous Maya villages, receiving consent from the people in the villages to investigate the local plants and traditional knowledge of plant medicines.

Such careful planning to make sure that all parties are on the same page is normal protocol for any ICBG grant program. The ICBG emphasizes cross-cultural partnership building through prior informed consent processes and benefits-sharing agreements. The source country, or the country in which scientists plan to conduct their research, can benefit through advance payments (such as milestone payments), royalties on commercial products, and training and provision of resources. In return, local communities agree to allow researchers access to what the ICBG defines as priority areas—places containing largely unexplored native plants and remedies—from which the information that is derived is used for drug development by a commercial partner. In some instances, such as the Maya ICBG, researchers may focus on species that are statistically reported as being used most often; many of these are cosmopolitan weedy species or cultivated species.

ICBG grants typically entail partnerships between university research groups, pharmaceutical companies, local scientists, and local indigenous communities. A number of ICBGs, however, have worked only on state-controlled lands, such as parks, and in the absence of traditional knowledge, and while having a company participant is encouraged some projects never engage a private partner. At the core of ICBG grants is academic partnership, in some cases entailing cooperative efforts with host country, government-run institutes. In cases in which indigenous peoples and their traditional knowledge are accessed, however, prior informed consent is a central element. If the indigenous tribes do not consent to the research, to sharing their knowledge of plants and medicines, the project may never get off the ground.

The ICBG Maya project embodied all the sought-after improvements in informed consent, equitable benefits sharing, and protection of indigenous knowledge. But within months of the grant being awarded, signs of trouble in Chiapas began to emerge. The Council of Indigenous Traditional Midwives and Healers of Chiapas (COMPITCH) had plans to commercialize herbal preparations based on local knowledge. When COMPITCH learned of the intentions of the Maya ICBG, it protested the project, claiming that it was an attempt by the United States to strip away the traditional knowledge of the Maya. The group insisted that all Maya villages cut off communication with the researchers. Nongovernmental organizations supporting the Maya, namely RAFI (now ETC Group), claimed that the prior informed consent process was deceptive and misleading.

At the time, the state of Chiapas was under occupation by Mexican troops, who were supposedly searching out members of the Zapatista National Liberation Army. The Zapatista consisted of destitute locals who were fed up with the government's indifference to the rights of the indigenous peoples of Chiapas and to the region's suffering economy. The merging of perceptions of bioprospecting with perceptions of oppression, as well as skepticism over the patenting of natural products, fueled emotionally charged accusations against US researchers, which were highly publicized in the media.

COMPITCH argued that only a small percentage of Maya communities had given consent for research but that in order for the project to be conducted, consent was needed from all Maya-speaking peoples. COMPITCH, however, lacked the authority to issue such a mandate, and hence the notion that universal consent was necessary was flawed. The argument, rather, was a tactical one, used to manipulate perceptions and disgrace the credibility of bioprospecting. Nevertheless, it made it impossible for the project to advance. The voice of the Maya communities who endorsed the project was suppressed by the turmoil raised by COMPITCH. Their story was never heard on an international level.

University of Georgia professor of anthropology and head of the
Maya project Brent Berlin had spent many years among the Maya
communities. He cared very much about the Maya and was dedi-
cated to rescuing their disappearing knowledge and to protecting
their ecosystems, which were deteriorating. He was troubled by the
sudden antagonism facing the project. The Maya bioprospecting
endeavor was the most transparent large-scale attempt at community-based natural-products drug discovery. It was the openness
with which the project was organized and presented to the Maya
and the opportunity for benefits sharing that had initially secured
the cooperation of locals. But after COMPITCH and RAFI stirred
up controversy, even the Mexican government, which had initially
supported the project, shied away. The situation in Chiapas was
precarious, and if the government were to back the project, there
was fear of a repeat of the 1994 Zapatista National Liberation Army
uprising, in which members of the group antagonized locals and the
international community, and in a few cases were reported to have
violently attacked government-run organizations.

The Mexican government withdrew its support, and shortly thereafter, ECOSUR pulled out. There was simply too much at stake.
Without the involvement of their Chiapas-based partner, the funding agencies, NIH and NSF, had no choice but to cancel the effort,
which they did in a formal announcement in 2001.

The ICBG program has supported multiple other bioprospecting projects, including efforts in Argentina, Cameroon, Chile and
northern Mexico, Costa Rica, Fiji, Indonesia, Madagascar, Nige-
ria, Panama, Papua New Guinea, Peru, the Philippines, Suriname,
Vietnam and Laos, and Uzebekistan and Kyrgyzstan. Though there
have been a few bumps from nongovernmental organizations along
the way, the majority of these projects have run smoothly, and each
has taken into careful consideration how best to engage in prior
informed consent and equitable benefits sharing.

The Panama ICBG program is working to create a framework
within the country that will enable local scientists to continue the
pursuit of drug discovery once the project has ended. This is made
possible in large part through technology transfer and training
opportunities. All participating parties and the Panamanian government will receive royalties from any commercial drugs that are

developed. The government intends to invest these profits in conservation and habitat protection. The work of training and infrastructure building in Panama also provides benefits beyond royalties, by securing eventual autonomy for its scientists, enabling them to sustain the discovery process, which in turn can fuel economic development.

Part of the Panama project included research at Coiba Island, which rests just off the country's Pacific coast. The ICBG-sponsored work at Coiba has been among the most successful efforts so far. The Coiba research led William Gerwick, a scientist at the Scripps Institution of Oceanography at the University of California, San Diego, to discover a new anticancer compound, named coibamide, in marine algae. The ICBG program and Gerwick's work helped to improve Panama's research infrastructure and economy. Some ninety scientists from Panama and the United States were trained for the program, and local lab technicians were hired to assist with the research. The program was a major factor in helping Coiba earn status as a UNESCO World Heritage site.

The ICBG Madagascar project is similarly focused on the discovery of new compounds for commercial development and on conservation and scientific and economic development within the source country. Botanists and marine biologists from three Malagasy groups—Madagascar's Centre National d'Application et des Recherches Pharmaceutiques (CNARP), the National Center for Oceanographic Research (CNRO), and the National Center for Environmental Research (CNRE)—were involved in gathering and testing plant and marine materials. Malagasy scientists, trained at US–based Dow AgroSciences, were also involved in exploring natural substances from Madagascar's microorganisms. CNARP, Dow, the Virginia Polytechnic Institute and State University, and the US company Eisai Research Institute were involved in the screening and investigation of isolated natural compounds to identify drug candidates. Madagascar was signed on to receive royalties on any commercial products spun out of the project.

In the process of identifying plants and other organisms as sources of potential drug compounds, scientists are also cataloging Madagascar's biodiversity, an effort fundamental to establishing effective conservation programs. These projects demonstrate initiative in overcoming the problems inherent in patent systems. The ICBG and the researchers it supports have taken it upon themselves to improve bioprospecting and have raised the bar for pharmaceutical companies and other entities that wish to work independently in natural-products discovery. In doing so, they have sent a strong message about respecting the traditional knowledge and cultural heritage of indigenous peoples. The ICBG has shown that cultural respect can go a long way.

While much work has gone into overcoming the ethical and practical problems associated with bioprospecting, there has been little effort to quench the abuse of natural products sold as dietary supplements. Over-the-counter herbal products are the worst offenders in the exploitation of traditional knowledge and plant medicines. Companies manufacturing these products capitalize not only on indigenous knowledge but also on raw scientific research.

Nonprescription products frequently are either purified compounds or structurally modified compounds. In some instances, however, structural modification reflects a deliberate attempt to evade regulatory processes, and in other cases it may represent a cost shortcut. Modified compounds can pose serious risks to health. A slight change in chemical structure can render an effective substance ineffective or even toxic. In contrast to the pharmaceutical industry, there are no formal procedures in place that require modified chemicals intended for over-the-counter use to be tested clinically for safety and effectiveness.

In 1994 the US government passed the Dietary Supplement Health and Education Act (DSHEA), which placed the responsibility of ensuring supplement safety prior to marketing on the shoulders of product manufacturers. The act defined dietary supplements as foods rather than as drugs, even though these products blur the distinction between foods and drugs. A variety of dietary supplements contain substances that exert druglike effects on the body, which differ from the well-defined metabolic processes the body employs to break down carbohydrates, sugars, proteins, and other nutrient substances in foods. Dietary supplements are strictly defined as substances taken by mouth and include herbal and botanical remedies, vitamins, minerals, amino acids, enzymes, metabolites, and other biological substances. DSHEA encouraged manufacturers to make honest statements on product labels and required disclosure of nutritional information on the back of the product under "Supplement Facts" and "Other Ingredients." Labels on dietary supplements must declare that the product is *not* intended to cure, prevent, or treat disease.

In the United States, FDA-approved pharmaceuticals, which include generic drugs, are the only products for which manufacturers can legally declare medical applications, and then only for those applications that have been approved. If a label on a dietary supplement were to allude to medical applications, it would be considered an illegal drug. Supplement manufacturers are hassled by the FDA only when a product has been reported to cause adverse effects or is found to contain an unlisted pharmaceutical. The organization accumulates information on adverse effects from consumers, health-care services, and manufacturers (who are required to report information on adverse events to the FDA) and eventually removes products from the market that exceed risk thresholds. In essence, the system of safety assessment lies with consumers, who may or may not report an adverse event or who may not even know that it is their responsibility to do so.

In 2007, to guide manufacturers in a more ethically and scientifically sound direction, the FDA published Current Good Manufacturing Practices, which since have been extended to the international market to improve supplements intended for export and import. The organization also published a consumer brochure titled "Tips for the Savvy Supplement User: Making Informed Decisions and Evaluating Information." The DSHEA was enacted because Americans wanted the freedom to choose to put into their bodies whatever they wanted, regardless of safety or efficacy. But many people remain unaware of the potential dangers of taking dietary supplements and have no way of fairly assessing whether or not a supplement is benefiting or harming their bodies. "Quality guaranteed" graces product labels, but who performs quality assessment?

Standardization requirements do not exist in the manufacture of nonprescription herbals. Some people do not believe in standardization for natural products, arguing that such substances require very little "processing" and consequently variation within and between batches is anticipated. In nature multiple compounds within a plant act in concerted fashion to elicit an effect. The production of these substances fluctuates naturally, being influenced by environmental factors, such as drought and foraging by animals. If a manufacturer were to standardize how its product is made, variation would be muted and the naturalness of the product would be reduced. Furthermore, substances necessary for producing the perceived "all-natural" response might be stripped out, resulting in a product that no longer generates an effect or that evokes a different response altogether. It is difficult to form any logical basis for or against the standardization of supplements, since there is no evidence available to back either side of the argument. The degree to which the concentrations of active ingredients can vary within and between products can be substantial. A study of the active ingredient ginsenoside in ginseng produced by manufacturers in the United States revealed that few of the products were labeled correctly as to the actual concentration of ginsenoside they contained and that between ginseng products the concentration of the substance varied by as much as fifteen-fold.

Supplements are also notorious for their potential for containing unapproved substances. The adulteration of supplements can occur in different ways, with the most frequent being the addition of an impure substance or the removal of an active ingredient. A 1998 study of traditional Chinese herbal products imported from Asia and sold in California discovered undeclared substances, including heavy metals and pharmaceutical compounds, in some 32 percent of products investigated. About 9 percent were found to contain more than one contaminating substance.

A number of dietary supplements have been associated with problems ranging from allergic reactions to cancer to drug interactions with prescription medicines. Extracts of ginkgo, for example, and substances such as yohimbine, echinacea, and St. John's wort have been known to cause various types of allergic reactions. Extracts from calamus, senna, borage, and yerba mate have demonstrated a tendency for carcinogenesis, based on laboratory and animal studies. Still, sales of dietary supplements in the United States, China, and other countries have climbed dramatically since the mid-1990s. Somewhat lagging behind this increase have been efforts to develop methods to reliably detect synthetic drugs and other substances used to adulterate supplements. Some of the most effective methods tested to date rely on liquid chromatography and mass spectrometry techniques that detect ions produced by electrospray and that are used in conjunction with systems that enhance ion spectra for accurate and rapid detection. In testing these approaches, researchers have stumbled upon numerous adulterants. In one study, out of 105 dietary supplements investigated, some 35 turned up positive for adulteration with an undeclared synthetic drug.

The manufacture and marketing of an "all-natural" substance that is spiked with synthetic drugs but sold based on applications learned from systems of traditional medicine misrepresents indigenous knowledge and obscures scientific efforts to better understand natural products. In traditional medicine, there exists inevitable variation between preparations of plant extracts, and the applications of such extracts may not be validated by modern science. Indigenous healers tend to be very selective about the plants they harvest for medicinal purposes—for example, only collecting mature plants or the minimal number of leaves necessary.

The appreciation for the natural world that is found within indigenous communities has become a subject of careful research in ethnobotany and ethnopharmacology. Combined with efforts to catalog Earth's myriad life-forms, these fields of research stand to bring indigenous knowledge and plants to the forefront of modern science in a way that respects cultural heritage and biodiversity while delivering new medicines to treat some of our most deadly diseases. While many manufacturers of dietary supplements may embrace these efforts, those who do not only encourage people to put into their bodies substances that are potentially harmful and in all likelihood unnecessary. Many people do not need to take supplements, but they do so anyway. The reasons for this are too many to generalize, yet it is difficult to ignore the role of the modern consumer mentality, which drives people to purchase many things they don't need. In buying and using supplements, however, people not only put their health at risk but also perpetuate the cycle of adulteration of unapproved plant-based products and exploitation of traditional knowledge.

It takes a great many species being harvested and sold every year to generate the economic significance that wild medicinal plants have maintained in recent years. But at large-scale commercial levels, particularly those demanded by the supplement trade, no single species of plant can sustain harvesting in the wild for any length of time.

In Canadian forests, wild populations of American ginseng are small, certainly not large enough to support harvesting. The viability of these *Panax quinquefolius* populations hovers at around 170 individual plants, and only a few populations are made up of more than this number. Once populations sink to fewer than 170 members, they become highly vulnerable to rapid decline. Once the population dwindles to between 30 and 90 individuals, it is extremely difficult for the species to rebound and expand again.

The limitations imposed on the harvesting of wild American ginseng are an effect of environmental stochasticity, or the difference between birth rate and death rate in a wild population from season to season as a result of unpredictable factors, such as local variations in weather patterns or overharvesting. Overharvesting is damaging to slow-growing species that are uniquely adapted to their habitats, particularly when these species are targeted for their roots or bark.

African cherry, which was added to the IUCN Red List in the late 1990s as a vulnerable species, was overharvested for timber and medicinal products. *Prunus africana* is an important economic resource for local communities in sub-Saharan Africa. It grows in the montane mixed forests of Cameroon, the Democratic Republic of the Congo, and Kenya and is also found on Madagascar. The tree is an important component of the ecosystem in which it grows— for example, serving as a source of food and habitat for various mammals. In 1995, because of its severe exploitation, primarily for timber, restrictions were placed on its trade. In recent years, international organizations have worked to develop sustainable-harvest guidelines. Permits and quotas on harvesting were introduced and adjusted for various purposes, such as large-quantity demands for timber and small-quantity uses for medicinal products. In the process of developing these guidelines, the International Standard for Sustainable Wild Collection of Medicinal and Aromatic Plants (ISSC-MAP) was considered, specifically to identify high-priority tasks needed for sustainable harvest and instances in which compliance was problematic.

American ginseng (*Panax quinquefolius*). (Photo credit: Dan J. Pittillo/ USFWS)

The overharvesting of medicinal plants is worrisome because it not only pushes species to the brink of extinction but also is a manifestation of human overconsumption. The problem of overharvesting was so significant by the end of the twentieth century that in 1994 the Medicinal Plant Specialist Group (MPSG) was established as part of the Species Survival Commission of IUCN. The MPSG was charged with establishing international standards for the harvesting of medicinal and aromatic plants in the wild to provide a broader understanding of how species can be harvested in a sustainable manner. These guidelines were eventually compiled, forming the ISSC-MAP.

Overharvesting, though perhaps driven on national or international levels, is a problem that must be confronted in local communities. Much the same way that ICBG projects are aimed at building relationships between indigenous peoples and researchers to facilitate drug discovery, conservation, prior informed consent, and benefits sharing, the inhabitants of local communities and government and global organizations must also find ways to work together to ensure the sustainable harvest of local resources. Commercial entities looking to make money quickly by rapidly harvesting plants in the absence of understanding environmental consequences and plant growth and reproduction are inflicting severe harm on nature. They often also are exploiting traditional knowledge and stealing away resources from rural communities.

In the quest for plant-based natural-products drug discovery, patience and an appreciation for nature and for traditional knowledge are paramount. These elements, however, are extremely difficult to realize in the modern era. Humankind is so disconnected from nature that many people struggle to understand why things such as biopiracy and the overharvesting of plants in the wild are unsettling. Learning to coexist with other human societies and with nature is of utmost importance and is perhaps among the greatest challenges faced by bioprospecting endeavors.

Underlying the success of natural-products drug discovery and the continued functioning of ecosystem services, which support the well-being of humanity worldwide, is the conservation of biodiversity. In order for us to truly understand the significance and the beauty of Earth's biodiversity, we must have opportunities to experience it firsthand, to step foot in the world's wild places, to see and feel nature as it exists isolated from the pressures of ever-sprawling human societies. The setting aside of protected areas, such as natural parks, natural reserves, and protected wildernesses, enables us to explore nature. Exploration helps cultivate a better understanding of the nuances that define interactions between different species of animals and between plants and animals. Such adventures into nature, however, require that large areas of Earth's ecosystems remain intact, which means cultivating within ourselves, and our future generations, a deep respect for nature. We also must better understand how we fit into the big picture of ecosystems.

8

The Forest for Its Trees

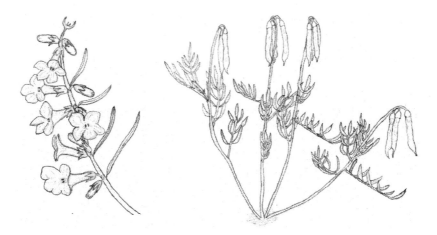

Penland beardtongue (*Penstemon penlandii*; left) and Osterhout milk-vetch (*Astragalus osterhoutii*), endangered species of Colorado.

BIODIVERSITY CONSERVATION REQUIRES that we understand the big picture of life on Earth, especially how our actions affect ecosystems, which contain all the living forms that make our existence possible. We, like all the creatures around us, are the products of evolution, born from nature, from generations spent among plants and animals. Our ancestors depended on nature for survival, and so do we, for everything from food to shelter to sustaining global economy. But there are so many more humans on Earth now that our demands on ecosystem services are outpacing their biological productivity. Clean air and water, carbon storage, and the existence of species of medicinal and agricultural value are rapidly becoming privileges, exceptions to ordinary services. In separating ourselves from the environment with concrete sidewalks, paved roads, and steel buildings, we have created not only a physical division between us and nature but a psychological one as well. The separation transcends industrialized societies worldwide, travels up the chain from individuals to communities to national governments. The notion that nature is dispensable has been cultivated through a gradual process, beginning with inventions that made human lives more comfortable and expressing itself now in the form of consumerism. The public perception of nature has been especially confused by politics and political agendas often aimed not at protecting natural resources or using them sustainably, but at profiting from them, often by whatever means necessary.

Working around misguided politics in biodiversity conservation is one of the challenges faced by the multiple organizations and institutes that have stepped up to lead conservation efforts. In finding ways to harness the natural value of places such as biodiversity hotspots, governing bodies have needed to sever themselves from traditional notions of nature's economic value, embodied in the cutting down of trees and the building of dams or harvesting of medicinal plants until an ecosystem is destroyed beyond recovery. Rather, to reap the greatest benefits of the most biologically diverse regions of the world, it has become understood that these areas must be protected, through conservation and preservation. Some terrestrial and aquatic areas deemed vulnerable to species loss or considered of extraordinary value to humankind have been embraced in the protection of national parks and similar entities. In other instances, prior informed consent and benefits-sharing projects, whether for timber, foods, or medicines, have encouraged the sustainable use of resources.

The ecosystem services of biodiverse lands are virtually impossible to measure in terms of gross domestic product. Instead, the well-being of Earth's biodiversity, and its function and ability to deliver services, is better reflected in the state of species populations. The Living Planet Index (LPI), which tracks changes in biodiversity among different groups of vertebrate species over time, is one measure used to assess the health of habitats. According to the

LPI, from 1970 to 2005 there was a 30 percent drop in terrestrial, freshwater, and marine species. Because animals depend on plants, the identification of plants and regions with large numbers of plants susceptible to loss is a major factor in conservation. Knowing which plants are threatened and which plant areas are home to threatened species of animals supports biodiversity conservation.

The Convention on Biological Diversity (CBD) has been an important player in initiating action toward plant conservation. The international treaty, implemented in 1993, binds its signatories to the conservation of biodiverse regions, allowing for the sustainable use of biologic and genetic resources. The treaty was fundamental in the movement toward benefits-sharing agreements. But similar to other international treaties that set guidelines on how resources are to be used and shared, the CBD has a relatively passive role in enforcing legally binding contracts. It serves instead as a framework for conservation that participating countries are obliged to incorporate into their own national legislation. But unlike conventions, such as the United Nations Convention on the Law of the Sea, CBD recognizes the fact that many developing countries cannot afford to financially support biodiversity conservation. CBD has its own financial portal, which it uses to communicate information about funding opportunities and cooperative projects, allowing less industrialized countries, such as the Democratic Republic of Congo, Gabon, and Papua New Guinea, to participate in conservation. These places are home to some of the world's most biologically productive habitats, and invoking various funding mechanisms for their protection is important to international conservation efforts.

In 2002 the CBD endorsed the Global Strategy for Plant Conservation (GSPC), which is meant to prevent the extinction of plants and promote benefits sharing and sustainable plant use. The GSPC is championed by 180 countries and is further supported by the Global Partnership for Plant Conservation (GPPC), an alliance of organizations whose mission is to protect biodiversity. One of GPPC's central goals is to implement the targets laid out by the GSPC. The global strategy consists of sixteen targets, which center on understanding, documenting, and conserving plant diversity; promoting the sustainable use of plant diversity; and providing the means, through education and training, for obtaining these goals. Participating countries were scheduled to meet these targets for 2010, but multiple challenges delayed progress, and plans were laid to extend the project through 2020. Some of the problems encountered included lack of financial support, lack of technological resources, and absence of organized political support. Such problems have been the bane of conservation for many decades, and overcoming them has been and continues to be a difficult process.

One of the major targets, the development of a "working list" of plants, known formally as *The Plant List*, saw substantial progress between 2002 and 2009 and was eventually published in late 2010. The list is essentially a centralized compilation of all known plants that includes every Latin name used for each species. The information is fundamental in the identification of plants for protection strategies and sustainable uses. The working list also supports the other targets of the GSPC, specifically the long-term goal of conserving the two-thirds of the world's plants that are expected to fall into the clutches of a threatened existence by the end of the twenty-first century. Botanists and conservationists worldwide are able to access the working list for their research.

Another goal of the GSPC is the identification of Important Plant Areas, or IPAs. The Important Plant Areas program was developed by Plantlife International. Plantlife defines IPAs as natural or mostly natural sites with significant botanical diversity or an important collection of rare or native plants or other vegetation of exceptional value. These areas constitute a fundamental part of an ecological system. More than thirty countries have identified IPAs and initiated work toward their protection. The caveat of IPAs, however, is that they are not necessarily conserved regions. They frequently are only areas identified as containing a diverse group of plants of economic value. The GSPC encourages countries to take the extra steps toward identifying ways in which IPAs can be used to feed into conservation. Getting the attention of people in local communities by attracting interest in scientific research and finding ways to tie conservation to income is cited by GSPC as one of the most effective approaches to ensuring IPA protection.

One of the interesting elements of the global plant conservation program and the identification of IPAs is the number of national botanic gardens and parks involved. These are places that the general public can visit to learn about native plants and animals and the role they play in local ecosystems. Translating the work of researchers worldwide to local levels creates increased tangibility for the average person. The knowledge we take away from botanic gardens can be incorporated into our lives, influencing the way we design and tend to our own gardens and pursue our individual interests in nature and conservation. National parks and botanic institutes and gardens have become important channels for conveying information to the public on how human activities affect the environment.

But learning how our actions impact nature is only part of the conservation puzzle. It is important that we understand the relationships between our activities and the environment's health, but it is equally important that we establish individual and community connections with nature. National parks and botanic gardens receive millions of visitors each year, creating countless opportunities for increasing conservation awareness. Kings Park and Botanic Garden in western Australia, for example, receives some 6 million visitors each year and includes supporting nature conservation among its goals. Visitors can connect with nature through educational services, hands-on activities, and guided walks. The garden supports a significant amount of plant research, including exploration of plant physiology and restoration ecophysiology (when native plants are reintroduced to partially or fully restored habitats), the study of recovery and storage of seeds from rare species, and genetic analyses of these species to better understand how they can be cultivated by human hands. The park also emphasizes the need to protect indigenous knowledge and cultural heritage, in the context of nature conservation.

The Royal Botanic Gardens, Kew, receives around 1.3 million visitors each year. Kew has top-notch scientists stationed all over the world. Working with the Chinese Academy of Sciences, Kew contributed to the establishment of the Southwest China Germplasm Bank of Wild Species, one of the largest wild plant conservation programs in the world. An agreement between Kew's Millennium Seed Bank Project and China's Germplasm Bank has promised the exchange and storage of several thousand seeds from China. Kew's Millennium Seed Bank, initiated in 2000, supports the collection and storage of seeds of wild plants at risk of extinction. At the end of 2009, seeds from nearly 10 percent of wild plants in the world had been stored.

Establishing the economic contributions of nature in global trade schemes has become fundamental to conservation. To this purpose, the Convention on International Trade in Endangered Species of Wild Fauna and Flora (CITES) developed strategies to prevent the further overharvesting and unsustainable exploitation of wild species as a result of international trade. CITES has been around for a while, having come into effect in 1975, after more than a decade of drafting. Member states must develop their own approach to implementing laws within the CITES framework. Meanwhile, CITES maintains a list of endangered species exploited by international trade and follows up with countries on legislative progress toward regulating the export and import of listed species.

In 1995 the African cherry was added to the CITES list of overexploited plants. It was added as an Appendix II species, the midlevel list, containing species that are threatened due to trade but are neither the most endangered (Appendix I) nor the least threatened (Appendix III). African cherry can be traded as long as the survival of the species in the wild is not compromised. The tree may be exported if a country has received a permit to do so, and with the exception of certain instances, a permit is not needed for its import. The plant's status in the Democratic Republic of the Congo, Equatorial Guinea, Burundi, Cameroon, Madagascar, Kenya, and Tanzania is routinely assessed to check CITES compliance and to determine whether the species' appendix ranking should be altered.

Each country that participates in CITES has at least one designated scientific authority and one management authority. In the United Kingdom, Kew serves as the scientific authority for CITES, investigating exploited plants, providing protection guidelines for plants to CITES, and working with UK enforcement agencies to ensure compliance. In the United States, CITES is channeled into the Endangered Species Act (ESA), which was passed in 1973 and now is jointly headed by the Fish and Wildlife Service (FWS) and the National Oceanic and Atmospheric Administration (NOAA) Fisheries Service. The ESA is a reflection of an optimism that has persisted in the United States, particularly ever since people became enamored with the Western Frontier and national parks. It ensures that our most fragile and threatened species will be protected from afflictions such as habitat loss and hunting. It also serves as the primary vehicle through which the country is able to remain in compliance with multiple international conservation agreements, including migratory bird treaties with Canada, Mexico, and Japan; the Convention on Nature Protection and Wildlife Preservation in the Western Hemisphere; the International Convention for the Northwest Atlantic Fisheries; and the International Convention for the High Seas Fisheries of the North Pacific Ocean. On a national level, through the ESA, states and conservation organizations may be eligible for federal financial support or be offered various incentives to develop and manage conservation programs that meet national and international expectations. According to the ESA, encouraging the development of conservation projects "is a key to meeting the Nation's international commitments and to better safeguarding, for the benefit of all citizens, the Nation's heritage in fish, wildlife, and plants."

The bald eagle, the Virginia northern flying squirrel, the brown pelican, Robbins' cinquefoil, and more than thirty other species have been recovered and delisted from the ESA. Sadly, some of the species first listed, including the Santa Barbara song sparrow, the dusky seaside sparrow, the Mariana mallard, and several species of fish went extinct, despite efforts to prevent their loss. Outright violation of the ESA is a major contributing factor to the ultimate downfall of species, with hunting, harvesting, and other harmful activities being pursued right in the backyards of FWS and law enforcement agencies.

In some instances, despite the ESA's good intentions, too many compromises are made on the welfare of habitats so as to accommodate land use demands. In fact, initiatives such as the ESA and CITES have been criticized for their limited scope, since they identify only single species for protection, rather than whole ecosystems. The ESA, however, includes a clause specifically referencing the importance of ecosystems to the survival of threatened species (found in Section 2, "Findings, Purposes, and Policies").

Still, there is much red tape suffocating ecosystems conservation in the United States. Disagreements over rights to land use, including the diversion of water to farms from wetlands and the grazing of sheep and cattle on prairies, where fragile soil and plant systems exist, create many bumps and rough relations along the way. Red tape is why, in countries where there exists legislative regulation to limit the human impact on the survival of plants and animals in the wild, species still are able to fall through the cracks and become lost to extinction. The ESA and CITES, which try to protect the world's most threatened species, are of very high importance. However, the long-term survival of endangered species will never be guaranteed in the absence of projects that address the underlying problems of habitat loss and fragmentation.

Penland beardtongue, a member of the plantain family, has been reduced to a small, isolated location along Muddy Creek in the northern region of Colorado. Its habitat measures a mere 0.5 by 1.5 miles. But *Penstemon penlandii* is a rugged little plant, a resilience masked by its delicate purple and blue flowers. *Penstemon* is known for its medicinal properties, with some species having been used by Native Americans. Several bioactive substances have been isolated from *Penstemon* species, making Penland beardtongue potentially valuable for drug discovery.

A remarkable characteristic of Penland is its exceptional adaptation to its tiny homeland. It grows at elevations of around 7,500 to 7,700 feet, and it is one of few species able to thrive in the area's selenium-rich clay soil. Its neighbors include scrubby species of sagebrush and rabbitbrush and resilient snowberry and winterfat. But Penland beardtongue grows with such sparseness within its habitat that it is completely defenseless against anything that threatens the integrity of the soil surface.

In 1989 Penland beardtongue joined the ESA's list. With such a small habitat, it would seem that few resources needed to be tapped to establish a successful conservation project. The work of researchers associated with FWS and the Denver Botanic Gardens has helped in this respect. But Penland is threatened by two thorns that long ago lodged themselves in the side of conservation—mining and off-road vehicles. Imposing restrictions on these activities has always been a difficult, if not impossible, task. Limiting the use of off-road vehicles, such as ATVs and snowmobiles, in wilderness areas is a constant battle between recreation seekers and conservationists. Erosion leading to habitat deterioration, the introduction of plant materials from invasive species attached to tires, noise produced by

engines, pollution emitted in exhaust fumes, and the lack of appropriate trail use and trail ethics are serious concerns associated with the
vehicles. Faced with these issues, plants like Penland beardtongue are
on the fast track to extinction. Indeed, Penland is not alone in its
struggle to maintain an existence along Muddy Creek. In the 1989
report leading to its listing as endangered, it was accompanied by
Osterhout milk-vetch (*Astragalus osterhoutii*).

Differences in opinion about how land should be used and dis-
regard for the sensitive nature of endangered species are issues that
need to be addressed on political and individual levels worldwide.
And while Penland beardtongue and Osterhout milk-vetch are bat-
tling ATVs, other plants find themselves in situations similar to that
of African cherry, near to extinction in the fight against overzealous
trade. In addition to African cherry, *Hoodia*, American ginseng, and
Asian ginseng all remained CITES Appendix II species through the
early twenty-first century. Yet, wild members of these species continued to be subjected to overharvesting and exploitation, activities
driven primarily by the high prices and quick profits of illegal trade.

Countries such as India and China, which are home to the vast
majority of plants known to possess medicinal value, experience
extensive problems with the trade of CITES-listed plants. Himalayan
mayapple is used in both Indian and Chinese systems of medicine for
a sundry of conditions, but it is found only at high, alpine elevations
in the western Himalayas, where it is known for the beautiful flowers
it produces each May. Interest in *Podophyllum hexandrum* for drug-
discovery purposes arose in the 1960s, when the roots and rhizomes
were discovered to contain a bioactive substance known as podophyllotoxin. Podophyllotoxin is the parent compound of the anticancer
drugs etoposide and teniposide, which are used in the treatment of
lymphoma and leukemia and testicular and lung cancers. Himalayan
mayapple is superior when it comes to the biological synthesis of
podophyllotoxin, producing somewhere around seventeen times the
amount synthesized by other *Podophyllum* species. The Himalayan
species has served as the primary source of commercial podophyllotoxin, but in recent years populations of the plant have fallen into
steep decline, mainly because of overharvesting for local and commercial uses. In 1990 the Himalayan mayapple was added to CITES
Appendix II, with its status unchanged following reevaluation in 2007.

Plant breeding to establish cultivated populations of Himalayan mayapple has met with little success. Methods for the total synthesis of podophyllotoxin, from which the active anticancer compounds can be derived, as well as generation of the toxin using engineered organisms, represent the primary alternatives to harvesting in the wild. Chemical synthesis techniques have been developed but are relatively inefficient. These approaches require further refinement before they can be used to meet commercial production demands. Other species of *Podophyllum* have been investigated for their potential role in the generation of anticancer agents from podophyllotoxin. Their meager production, however, meets with similar problems associated with the inability to fulfill commercial demand. It may take a combination of chemical synthesis and derivation from other, less-threatened *Podophyllum* species to ultimately relieve commercial harvesting pressures placed on the wild Himalayan mayapple. In the meantime, countries must be more attentive to restrictions on the plant's international trade, and communities in the western Himalayas must find ways to prevent further human encroachment and destruction of medicinal plant habitat.

In 1995 red sandalwood, also known as red sanders, joined CITES Appendix II, having been listed because of exploitation for timber and medicinal and dye products. A decade and a half later, *Pterocarpus santalinus* remained an Appendix II species and was listed as endangered by the IUCN. The species is endemic to a small area in the southern portion of the Eastern Ghats mountain range in India, where it prefers the soils of dry deciduous forest. It has long been prized for the furniture and instruments that are fashioned from its heartwood. The red dye produced from the wood is called santalin and is used in Hindu ceremonies and as a histological stain. In medicine, the tree's wood is used in the preparation of concoctions to treat fever, skin conditions, dysentery, headache, ulcers, and blood and liver diseases. Its bark is used to treat diabetes-like symptoms and microbial infection. Studies of red sandalwood have indicated that bark extracts are able to lower blood glucose levels, so the plant now represents a promising source for pharmaceutical research of new antidiabetes agents. Demand for red sandalwood is extremely high, both inside and outside India. In addition to timber, medicines, and dyes, it is used to produce sandalwood oil and incense. Trade restrictions and the lack of consistency in legislation concerning the tree's trade have led to extensive smuggling of wood and restricted products. The tree can be cultivated, but it is a slow-growing species, taking several decades to reach maturity. If red sandalwood is to survive in the wild, tighter local enforcement of trade regulations is needed.

Although international conventions have proven valuable in prodding countries down the path to biodiversity conservation, because they are only frameworks and are unable to do much beyond giving the occasional slap on the hand to recalcitrant countries they are able to only partially fulfill the conservation targets to which they aspire. The utility of international conventions on conservation is often brought into question, since countries inclined to engage in conservation will do so in the absence of international treaties, just as countries inclined to exploit species and habitats will do so regardless of the regulations suggested by international conventions. But this does not diminish their importance. The guidelines and leadership provided, and in the case of the CBD the funding opportunities presented, enable change for the better.

International and national legislation can go a long way toward changing the world. The ultimate success of enacting eco-friendly regulations, however, depends very much on the voices of individuals in communities. Excellent models for illustrating the far-reaching impacts of the collective action of citizens are smoking bans. In the latter part of the twentieth century, as the negative health effects of secondhand smoke were exposed, public health concerns over smoking in workplaces and public places increased dramatically. Drunk driving is illegal because it places the welfare of others at stake, and so it is with smoking in public places. Cardiovascular diseases and respiratory conditions such as lung cancer caused by secondhand smoke are preventable, and in places that have implemented smoking bans, prevention is paying off. Towns and cities where bans exist have experienced significant decreases in the incidence of acute heart attacks. On average, the incidence of heart attacks in the places investigated decreased by 15 percent in the first year of the ban relative to years before the laws were enforced. Three years into the bans, a 36 percent decrease in incidence of heart attacks was detected. In the long run, smoking bans will save the world trillions of dollars in preventable health-care expenses.

If people were as passionate about the welfare of the environment and other species as they have been about the right to breathing smoke-free air in public spaces, the conservation of biodiversity hotspots would be in much better shape than it is currently. The magnitude of what we could accomplish by initiating conservation efforts on local levels, which feed into conservation trends on national and international levels, is enormous. The difference between environmental conservation and smoking bans, however, lies with the fact that all of us, in one way or another, contribute to the activities that are undermining the survival of species and ecosystems. As a result, the collective voice of individuals in communities is far quieter when it comes to supporting conservation than it is when it comes to lobbying on behalf of smoking bans. People generally are not willing to make lifestyle sacrifices to save little-known species like Penland beardtongue.

These days, skepticism about the climate and environment has become a worrisome threat to biodiversity conservation. But the evidence exists—increasing numbers of studies have concluded that the world is warming, its glaciers are melting, and its biodiversity is decreasing. Still, many people would rather point fingers or deny the situation rather than take responsibility for their actions. Within the public sphere, understanding of the scope of the issues faced by conservation is tangential. People may read about biodiversity in popular science magazines or other media, and they recycle and take reusable shopping bags to the grocery store. But unlike smoking, where there was broad recognition of its adverse effects on human health, the impact of biodiversity deterioration on our well-being is recognized by comparatively few. The lack of knowledge and public concern about what is happening to life on Earth as a result of our activities causes conservation efforts to limp along.

There are also major hindrances to prioritizing biodiverse areas for conservation. Examples include determining the size of land area that must be set aside, which generally must be very large to ensure that ecosystems can maintain their functions, and determining the value of these places in economic terms. Ecosystem services historically have been left out of economy and policy discussions because measuring their monetary value with any remote degree of accuracy was too difficult. In the past, many ecosystem services could exist outside the frame of economics, since they were so prevalent and their exhaustion through human activities was perceived as unlikely. But things are different now. There are so many people in the world, and the population is growing so quickly, that many ecosystems services, in order to last and sustain humanity, must be given economic meaning.

Since the latter part of the twentieth century, researchers have been devising increasingly accurate valuation methods for these oft-considered "abstract" services. If conservation efforts are to fit into the global picture, we must admit that it is no longer feasible to protect nature by relying solely on notions that nature is priceless and protecting it is the right thing to do. Many people may operate under these honorable guidelines, but something more is needed to support conservation on the scale on which it needs to happen. In some instances it is not immediately apparent what, exactly, is the morally right action to take when it comes to protecting nature. Is it right to restrict the harvesting of a threatened plant that is the source of a live-saving anticancer drug or a source of food for a growing community? As some researchers have already pointed out, in making decisions in such instances, we automatically place value on nature and human life. Since we are doing so already, despite the uncertainties that may exist, why not attempt to make more fully informed value assessments of ecosystem services? To do this, conservation economists must first assess all the individual services an ecosystem offers. Services can be broken down into functions, such as the availability of water for agriculture as a result of ecosystem water supply and regulation and the formation of soil from the weathering of rock and the consequent accumulation of organic materials.

Recreational, aesthetic, and scientific opportunities made available as a result of ecosystem services are also part of the valuation process. Like any other exercise in determining market and commercial value, terms such as supply, demand, cost, quantity, price, consumer surplus, and producer surplus are thrown about. Consumer surplus essentially means the benefits we gain from the exchange of goods, and producer surplus represents the benefits that the producer (in this instance, the ecosystem) derives as a result of producing the goods consumed. Based on the information available for each service within a specific ecosystem, and drawing on either known data or certain assumptions for demand, the value of the service can be determined. This can be done by considering only the benefits to the producer or the benefits to both producer and consumer. These measures can then be multiplied over the entire land area of ecosystems, enabling global value to be determined.

A study led by American ecological economist Robert Constanza that was published in 1997 in the journal *Nature* indicated that the value of services provided by all Earth's ecosystems combined averaged $33 trillion per year, with an average global gross national product of $18 trillion. These figures were minimum estimates. Accounting for inflation, the value of ecosystem services today is likely to be much higher.

Expecting people to suddenly think of ecosystem services in economic terms presents multiple problems, and in the wrong hands, knowledge of services value could be leveraged for profit in ways that only encourage continued destruction. We already are in serious debt to ecosystem services. We have permanently lost some services, such as the protection of soil surfaces from wind erosion provided by the once vast prairies of North America, now reduced to about 1 percent of their original size. We have no choice but to make up this debt, and soon. It is a matter of species survival, or at least of sustaining the continued survival of such large populations of *Homo sapiens*. If we do not make an effort to repay nature for all that she has given us, two or three generations down the road the world will be a miserable place for humans.

Because the issues associated with bringing the world in line with nature are so extraordinarily complex, encompassing not only considerations of the demands we place on ecosystem services but also the disparities in demand arising from differences in socioeconomic concerns between developed and developing countries, scientists are not holding their breath. While models such as DPSIR (driving forces, pressures, state, impacts, and responses) are being used to develop indicator frameworks that address the issues of societal pressures on the sustainable use of ecological resources, ecologists and botanists are busy collecting and characterizing animal and plant species new to science. Likewise, ethnobotanists and ethnopharmacologists are attempting to find new compounds to cure already troublesome diseases such as HIV/AIDS. But time is running short, and conservationists can use all the help they can get. As individuals, this is where our opportunity to influence the future of biodiversity presents itself.

Environmental and ecological stewardship—taking responsibility for the health of the environment—is one of the simplest ways each of us can make a difference. In the United States, the Environmental Protection Agency (EPA) promotes stewardship and provides a mechanism by which people across the country can learn about opportunities to become involved in conservation and environmental projects. EPA public awareness campaigns provide educational resources and open doors to public involvement in protecting ecosystem services. The little things we do in looking after the welfare of the environment add up such that, over time, as each of us becomes more environmentally efficient, community, state, and ultimately national demand on ecosystem services is reduced. The success of our efforts provides ecological security. Earth's unique habitats and ecosystems, from its rain forests to grasslands and oceans, will retain their functional, aesthetic, and scientific value for the benefit of our children and grandchildren.

Our experiences with nature shape our perception of the environment and determine our sense of attachment to it. Our innate attraction to nature, biophilia, may explain why we feel attached to particular outdoor environments and are drawn to and fascinated by animals and plants. But because our biophilic drive has weakened as we have become more technologically oriented, realizing our attachment to and our place within the natural world has become extraordinarily difficult.

In the process of bioprospecting and working with indigenous peoples, researchers have learned a great deal about place attachment. The sense of rootedness in nature is notably strong among indigenous groups. Their concern over finding ways to mesh aspects of industrialized life with ancient elements of their societies demonstrates the difficulty of synthesizing old and new while taking into consideration the need to retain a sensitivity to nature, traditional knowledge, and cultural heritage. Researchers working with indigenous tribes have themselves adopted a sense of attachment to the places these groups inhabit, illustrated by the ICBG Maya project discussed in chapter 7. Although pressures for the discovery of patentable bioactive compounds were present, Brent Berlin and other researchers shared the concerns raised by the Maya and were prepared to take the extra steps needed to respect the connection to nature and the knowledge and heritage embodied by the Maya.

Of all known tropical species, plants and animals, less than 1 percent have been investigated for potential drug development. As more species are studied, the number of new drugs entering clinical trials and receiving approval for marketing is likely to increase as well. Bioprospecting, however, must be tied to sustainable use and plant conservation. The discovery of new bioactive compounds has little meaning if the compound cannot be harvested sustainably or generated synthetically or semi-synthetically. The number of new plant compounds to be discovered is also capped by the number of different plant species available for investigation. Because species and biodiversity are threatened by human activities, the number of species that are available to science is now a function of conservation. Scientists have estimated that at the current rate at which plants and animals are going extinct, one new major drug is being lost every two years. This is significant for a variety of reasons, but particularly so in the context of recent declines in the number of new compounds making it into clinical trials and surviving approval processes.

In the 1990s, following the passage of a law in the United States requiring pharmaceutical companies to pay a fee to the FDA when a drug was submitted for approval, the organization suddenly had money to hire more reviewers to assist in the approval process. As a result, the process that had long been criticized for being too slow swung the other way, accelerating beyond expectation. In 1996, 53 new molecular entities were approved, more than double the number of previous years. Almost immediately concerns were raised about the quality of the agents entering the market. Not surprisingly, the trend did not last long. By 2002 the number had dropped to 17 approvals for new molecular entities, dipping down to 16 in 2007. While part of the reason for this decline was the sudden increase in the number of approved drugs later found to possess dangerous side effects, causing their recall and removal from the market and prompting a more rigorous approval process, it also was

a reflection of a decrease in quality of the new entities being submitted for approval. Novelty in drug development remains remarkably low. The sense that there has occurred a loss of meaning in
conventional medicine, combined with dissatisfaction in its ability
to meet health-care needs, further reinforces the value of exploring
traditional systems of medicine and herbal remedies for incorpora-
tion into Western practice.

For many agents that are approved, the clinical trials process is
generally so rigorous as to weed out the potentially harmful agents
or is at least unbiased enough to enable informed assessments to be
made in weighing benefits versus harms. Further, some of the most
recent drug recalls have occurred not because of a slip-up in the
FDA's approval process, but because of the submission of misleading information or the withholding of critical data. Pharmaceutical
companies often later make up for these deliberate approval side-
steps by paying for them in legal settlements. But this does little
to appease the wariness of medicine that such incidents generate
among the public. Our health is perhaps placed in even greater
danger by supplement companies that add pharmaceuticals to their
"natural" products, which can be purchased over-the-counter by
anyone. Despite the FDA's ability to recall adulterated products,
companies continue to manufacture and sell tainted supplements.
Other than paying out large sums of money for their wrongdoings,
there are few ways in which a company can be punished for deliber-
ately misleading consumers and placing their health at risk.

Most countries and unions made up of member states have their
own drug-approval regulations, and each must deal with recalls for
pharmaceuticals and supplements. People seem to be generally more
forgiving of supplement companies, perhaps because regulatory
standards are lower or perhaps because their recalls don't make it
into major news headlines. For pharmaceuticals, however, each time
that a marketed drug presumably meeting the quality standards of
national regulatory bodies is discovered to cause patients more harm
than good, there seems to follow an associated decrease in consumer
confidence in conventional medicine. The success of pharmaceutical regulatory processes relies on the products feeding into them
and on company transparency. Patents on reliable drugs supporting
the majority of revenues for some companies are expiring, and in

the meantime, the shortcuts taken in the pharmaceutical industry to shuffle new molecular entities into the hands of the FDA and other regulatory agencies for approval have resulted in the production of substandard agents.

One issue at the heart of all these problems is the depleted supply of parent compounds. Natural products fulfill a major role in modern medicine. Some 56 percent of the 150 best-selling prescription agents marketed in the United States are products of drug discovery in nature. These agents have an estimated value of $80 billion. Making an effort to venture into nature and to engage in prior informed consent and benefits-sharing drug discovery stands to generate a tremendous amount of revenue, for pharmaceutical companies and for the local communities in which the developed agents and the knowledge of applications of local medicinal plants are discovered.

At its most fundamental level, drug discovery is a very noble cause, with the intention of saving lives. But in the private sector, drug development rides the turbulent gamut of the market and is guided by the bottom line. The pursuit of revenue can very easily send companies down paths that distance them from the real reasons the generation of new medicines is so important. Drug discovery initiated in academia, though sometimes resulting in small, private spin-off companies, is by nature a very different beast than big pharma. At its core, academia is an expression of opportunity, freedom, and sharing, all the things that often contradict the pharmaceutical industry. The foremost difference between industry and academia is that the latter operates under an open-door policy. It is therefore much more amenable to benefits-sharing bioprospecting endeavors and to reinvesting the results of research in the sustainable use of plants and plant conservation.

There is no shortage of opportunities for natural-products drug discovery. In the epoch of ecosystem services and ecological economics, the investigation of large numbers of natural substances means potentially enormous payoffs and reinvestment in plant protection. But plants are not the only sources of natural products that stand to benefit from investigation for drug discovery. Microorganisms, amphibians, and a variety of marine life-forms also have found a place in the search for novel therapeutics. In the world's oceans, as the Coiba Island success discussed in chapter 7 shows, the discovery of medicinal compounds synthesized by marine organisms has opened doors for the development of new drugs and the conservation of ocean habitat. And similar to plants, the generation of synthetic and semi-synthetic methods for the production of bioactive substances from marine life plays an important role in precluding the need to harvest organisms from their ocean homes.

The number of living organisms from which drug discovery can benefit is astounding. But this number is decreasing at such a tremendous rate that if we do not act soon, the potential services these life-forms can offer us will disappear forever. Plants, animals, and the ecosystems they inhabit exist for reasons beyond simply serving humankind. They support one another, and the relationships shared between them are much more ancient than the relationships we share with them. In this respect, we are newcomers on Earth. Yet, we are the most highly advanced species in existence, at least in the way we view evolutionary progress. We have the gift of forethought, an element not wholly unique to humans but still of extraordinary importance, and we possess the ability to act on forethought. Prevention, touted as the key to good health, now forms the crux of human existence, determining the future state of our species on Earth. The deterioration of the ability of ecosystems to support our modern lifestyles not only strips us of water and food but also of the simplicity to appreciate nature. When the magnificence of the world's wild places is compromised for the sake of sprawling development and for our personal interests, and the compromise is explained away with words like "progress," "advancement," and "recreation," we might be in denial about the actual consequences of our actions or, in other instances, unaware of the consequences.

We need to address fundamental questions about the ways in which we use natural resources, particularly energy. Knowing where our energy comes from—whether nonrenewable oil, natural gas, and coal or renewable wind, solar, and waves—helps us understand the extent of our impacts on nature. The same can be said for knowing more about the sources of the foods we eat and understanding the impacts of modern transportation. Hundreds of vehicles carrying one person each, rather than a few vehicles carrying dozens of

Protected areas, such as Glacier National Park (Bowman Lake shown) in Montana, provide excellent opportunities to cultivate our appreciation of nature. (Photo credit: Jeremy D. Rogers)

people, are a reflection of consumerism. In ten or twenty years, the planet and its atmosphere will not be able to tolerate vehicle emissions for the billions of people globally who will be driving cars.

Becoming engaged with nature is vital in better understanding how our activities affect the environment and in achieving progress toward conservation. As individuals, we can help plants and animals by regularly disconnecting from our cell phones, televisions, and computers and stepping into the world around us. Stopping and thinking for a few moments about the world we inhabit can have a profound impact on our actions. And for some of us, this might even mean that we find ourselves reestablishing a meaningful appreciation for and connection with nature.

Biodiversity conservation originates with public interest and concern. Making efforts as individuals to lessen the burden we place on Earth is the first—and most significant—step toward saving plants and animals. Much is at stake. The drugs that are developed from natural-products discovery are just some of the many reasons why we must, as individuals and societies, make an effort to protect sensitive habitats. Just as smoking bans gained momentum through public awareness, so too does the advance of biodiversity conservation depend on public embrace.

Our actions will come full circle, whether positive or negative. Because conservation efforts are gaining support worldwide, and because scientists' knowledge of what needs to happen to help the environment is rapidly expanding, we should believe that change revolving around the needs and welfare of Earth's ecosystems is possible. In protecting plants from overharvesting and preventing the pirating of traditional knowledge, and in meandering through our gardens and exploring our national parks, we collectively are acting to conserve the natural wonders of the world. In doing so, we are saving plants that may be sources of new medicines, which we may someday find ourselves relying on to cure disease. Whatever the drugs may be, we will be relieved and proud that our actions contributed to the conservation of the biodiversity and cultural diversity underlying the discovery and development of such medicines. The magnificence and beauty of nature will prevail.

A walk along the rim of the Grand Canyon, across the mountains of Olympic National Park, or through the bushy undergrowth of Costa Rica's Monteverde cloud forest will have new meaning. We will know why our role in conservation is so important. The value of our relationship with nature, our exploration of it and our interactions with it, will unfold in vivid relief before us, revealing our true existence, born from the same, wild roots as all other life on Earth.

Bibliography

Achard, Frédéric, et al. "Determination of deforestation rates of the world's humid tropical forests." *Science* 297.5583 (2002): 999–1002.

Agrawal, A., and I. S. Fentiman. "NSAIDs and breast cancer: a possible prevention and treatment strategy." *International Journal of Clinical Practice* 62.3 (2008): 444–449.

Avancini, Graziela, et al. "Induction of pilocarpine formation in jaborandi leaves by salicylic acid and methyljasmonate." *Phytochemistry* 63.2 (2003): 171–175.

Barnes, Patricia M., et al. "Complementary and alternative medicine use among adults: United States, 2002." *Seminars in Integrative Medicine* 2.2 (2004): 54–71.

Bateman, Richard M., et al. "Early evolution of land plants: phylogeny, physiology, and ecology of the primary terrestrial radiation." *Annual Review of Ecology and Systematics* 29 (1998): 263–292.

Bell, Charles D., Douglas E. Soltis, and Pamela S. Soltis. "The age of the angiosperms: a molecular timescale without a clock." *Society for the Study of Evolution* 59.6 (2005): 1245–1258.

Berry, Susan, Steve Bradley, and Channel Four (Great Britain). *Plant Life: A Gardener's Guide*. London: Collins & Brown in association with Channel Four Television, 1993.

Bodeker, Gerard. "Traditional medical knowledge, intellectual property rights, and benefit-sharing." *Cardozo Journal of International and Company Law* 11 (2003–2004): 785–814.

Borchardt, John K. "The beginnings of drug therapy: ancient Mesopotamian medicine." *Drug News & Perspectives* 15.3 (2002): 187.

Bossuyt, Franky, et al. "Local endemism within the Western Ghats–Sri Lanka biodiversity hotspot." *Science* 306.5695 (2004): 479–481.

Bridges, E. M., and J. H. V. Van Baren. "Soil: an overlooked, undervalued and vital part of the human environment." *The Environmentalist* 17 (1997): 15–20.

Burns, William R. "East meets West: how China almost cured malaria." *Endeavour* 32.3 (2008): 101–106.

Carson, Walter, and Stefan Schnitzer, eds. *Tropical Forest Community Ecology.* Chichester, NH, and Malden, MA: Wiley-Blackwell, 2008.

Chen, Y., et al. "Determination of synthetic drugs used to adulterate botanical dietary supplements using QTRAP LC-MS/MS." *Food Additives and Contaminants* 26.5 (2009): 595–603.

Chivian, Eric, and Aaron Bernstein, eds. *Sustaining Life: How Human Health Depends on Biodiversity.* Oxford and New York: Oxford University Press, 2008.

Cochrane, Mark A., and Christopher P. Barber. "Climate change, human land use and future fires in the Amazon." *Global Change Biology* 15.3 (2009): 601–612.

Constanza, Robert, et al. "The value of the world's ecosystem services and natural capital." *Nature* 387 (15 May 1997): 253–260.

Council for Scientific and Industrial Research (CSIR). "The San and the CSIR announce a benefit-sharing agreement for potential anti-obesity drug." Press release, 24 March 2003.

Cousens, Rogers, Calvin Dytham, and Richard Law. *Dispersal in Plants: A Population Perspective.* Oxford: Oxford University Press, 2008.

Cowell, F. R. "Gardens as an art form." *British Journal of Aesthetics* 6.2 (1966): 111–122.

Daily, Gretchen, et al. "The value of nature and the nature of value." *Science* 289 (21 July 2000): 395–396.

Dalley, Stephanie. "Ancient Mesopotamian gardens and the identification of the Hanging Gardens of Babylon resolved." *Garden History* 21.1 (1993): 1–13.

Dalton, Rex. "Political uncertainty halts bioprospecting in Mexico." *Nature* 408.278 (16 Nov. 2000).

De Duve, Christian. "The origin of eukaryotes: a reappraisal." *Nature Reviews Genetics* 8 (May 2007): 395–403.

DeKosky, Steven T., et al. "*Ginkgo biloba* for prevention of dementia: a randomized controlled trial." *Journal of the American Medical Association* 300.19 (2008): 2253–2262.

Demaine, Linda J., and Aaron X. Fellmeth. "Patent law: natural substances and patentable inventions." *Science* 300.5624 (2003): 1375–1376.

Desai, Manoj C., and Samuel Chackalamannil. "Rediscovering the role of natural products in drug discovery." *Current Opinion in Drug Discovery & Development* 11.4 (2008): 436–437.

Desmet, P. G., et al. "Integrating biosystematic data into conservation planning: perspectives from southern Africa's Succulent Karoo." *Systematic Biology* 51.2 (2002): 317–330.

Desmond, Ray. *Sir Joseph Dalton Hooker: Traveller & Plant Collector.* Woodbridge, UK: Antique Collectors' Club with Royal Botanic Gardens, Kew, 1999.

Diamond, Jared. *Guns, Germs, and Steel.* New York: Zebra Bouquet, 1999.

Ding, G., et al. "Genetic diversity across natural populations of *Dendrobium officinale*, the endangered medicinal herb endemic to China, revealed by ISSR and RAPD markers." *Russian Journal of Genetics* 45.3 (2009): 327–334.

Drews, Jürgen, and Stefan Ryser. "The role of innovation in drug development." *Nature Biotechnology* 15 (1997): 1318–1319.

Driver, Amanda, et al. "Succulent Karoo Ecosystem Plan Biodiversity Component Technical Report." Cape Town, Cape Conservation Unit, Botanical Society of South Africa (2003).

Duffy, John F. "Rules and standards on the forefront of patentability." *William & Mary Law Review* 51.2 (2009): 609–653.

Duncan, Dayton, and Ken Burns. *The National Parks, America's Best Idea: An Illustrated History.* New York: Alfred A. Knopf, 2009.

Dyall, Sabrina D., Mark T. Brown, and Patricia J. Johnson. "Ancient invasions: from endosymbionts to organelles." *Science* 304.5668 (2004): 253–257.

Enserink, Martin. "Infectious diseases: source of new hope against malaria is in short supply." *Science* 307.5706 (2005): 33.

ETC Group. "Peruvian farmers and indigenous people denounce maca patents: extract of Andean root crop patented for 'natural Viagra' properties." *Genotype* 3 July 2002.

Ex parte Latimer, 12 March 1889 (C.D., 46 O.G. 1638), US Patent Office, *Decisions of the Commissioner of Patents and of the United States Courts in Patent Cases, 1889.* Washington, DC: US Government Printing Office, 1890, pp. 123–127.

Farrar, Linda. *Ancient Roman Gardens.* Stroud, UK: Sutton Publishing, 1998.

Federal Register. "Utility examination guidelines." 66.4 (2001).

Fleming, Alexander. "Correspondence: penicillin." *British Medical Journal* (13 Sept. 1941): 386.

Foster, Steven, and Rebecca L. Johnson. *National Geographic Desk Reference to Nature's Medicine.* Washington, DC: National Geographic, 2008.

Gaston, Kevin J. "Biodiversity and extinction: species and people." *Progress in Physical Geography* 29.2 (2005): 239–247.

Gindin, E. Jane. "Maca: traditional knowledge, new world." *The Trade & Environment Database.* American University, School of International Service, Washington, DC (2002).

Global Health Matters. "Panama prospecting—going for the green." 8.3 (May–June 2009).

Gole, Cheryl. *The Southwest Australia Ecoregion: Jewel of the Australian Continent.* Wembley, WA: Southwest Australia Ecoregion Initiative, 2006.

Goodman, Jordan, and Vivien Walsh. *The Story of Taxol: Nature and Politics in the Pursuit of an Anti-Cancer Drug.* Cambridge, UK, and New York: Cambridge University Press, 2001.

Gribbin, Mary, and John R. Gribbon. *Flower Hunters.* Oxford and New York: Oxford University Press, 2008.

Groombridge, Brian, and Martin D. Jenkins. *World Atlas of Biodiversity: Earth's Living Resources in the 21st Century.* Berkeley and London: University of California Press, 2002.

Gu, S., et al. "Isolation and characterization of microsatellite markers in *Dendrobium officinale*, an endangered herb endemic to China." *Molecular Ecology Notes* 7.6 (15 May 2007): 1166–1168.

Hahs, Amy K., et al. "A global synthesis of plant extinction rates in urban areas." *Ecology Letters* 12.11 (31 Aug. 2009): 1165–1173.

Hamilton, Alan, and Elizabeth Radford. *Identification and Conservation of Important Plant Areas for the Medicinal Plants in the Himalayas.* Plantlife International and the Ethnobotanical Society of Nepal, 2007.

Harvey, John H. "Vegetables in the Middle Ages." *Garden History* 12.2 (1984): 89–99.

Hileman, Bette. "Accounting for R&D: many doubt the $800 million pharmaceutical price tag." *Chemical and Engineering News* 84.25 (19 June 2005): 50.

Howell, Catherine H. *Flora Mirabilis: How Plants Have Shaped World Knowledge, Health, Wealth, and Beauty.* Washington, DC: National Geographic (2009).

Hubbard, Roderick E., ed. *Structure-Based Drug Discovery: An Overview.* Cambridge, UK: RSC Pub., 2006.

Hughes, Bethan. "2008 FDA drug approvals." *Nature Reviews Drug Discovery* 8 (Feb. 2009): 93–96.

Huxtable, Ryan J., and Stephan K. W. Schwarz. "The isolation of morphine—first principles in science and ethics." *Molecular Interventions* 1.4 (2001): 189–191.

International Union for Conservation of Nature and Natural Resources (IUCN). "Extinction crisis continues apace." 3 Nov. 2009.

Jackson, Jeremy B. C. "Ecological extinction and evolution in the brave new ocean." *PNAS* 105, suppl. 1 (2008): 11458–11465.

Janick, Jules. "Plant exploration: from Queen Hatshepsut to Sir Joseph Banks." *HortScience* 42.2 (2007): 191–196.

Jashemski, Wilhelmina F. *The gardens of Pompeii: Herculaneum and the villas destroyed by Vesuvius.* New Rochelle, NY: Caratzas Bros., 1979.

Jiang, Yuan, et al. "Impact of land use on plant biodiversity and measures for biodiversity conservation in the Loess Plateau in China—a case study in a hilly-gully region of the northern Loess Plateau." *Biodiversity and Conservation* 12.10 (2003): 2121–2133.

Kellert, Stephen R. *Kinship to Mastery: Biophilia in Human Evolution and Development.* Washington, DC: Island Press, 1997.

Kellert, Stephen R., and E. O. Wilson, eds. *The Biophilia Hypothesis.* Washington, DC: Island Press, 1993.

Kerr, Jeremy T., and David J. Currie. "Effects of human activity on global extinction risk." *Conservation Biology* 9.6 (1995): 1528–1538.

Kharkwal, A., et al. "Genetic variation within and among the populations of *Podophyllum hexandrum* Royle (Podophyllaceae) in western Himalaya." *PGR Newsletter* 156 (2008): 68–72.

Krief, Sabrina, Claude Marcel Hladik, and Claudie Haxaire. "Ethnomedicinal and bioactive properties of plants ingested by wild chimpanzees in Uganda." *Journal of Ethnopharmacology* 101 (2005): 1–15.

Kruczynski, Anna, and Bridget T. Hill. "Vinflunine, the latest *Vinca* alkaloid in clinical development: a review of its preclinical anticancer properties." *Critical Reviews in Oncology/Hematology* 40.2 (2001): 159–173.

Larsen H. O., C. S. Olsen, and T. E. Boon. "The non-timber forest policy process in Nepal: actors, objectives and power." *Forestry Nepal* 1.3/4 (2004): 267–281.

Larson, Richard S. (ed.). *Bioinformatics and Drug Discovery*. Totowa, NJ: Humana Press, 2005.

LEAD: The Livestock, Environment and Development Initiative. *Livestock's Long Shadow: Environmental Issues and Options*. Rome: FAO, 2006.

Ledford, Heidi. "Plant biology: the flower of seduction." *Nature* 445 (22 Feb. 2007): 816–817.

Leistner, Eckhard, and Christel Drewke. "*Ginkgo biloba* and ginkgotoxin." *Journal of Natural Products* 73.1 (2010): 86–92.

Lewis, Walter H., and Memory P. F. Elvin-Lewis. *Medical Botany: Plants Affecting Human Health*, 2nd ed. Hoboken, NJ: Wiley, 2003.

Lightwood, James M., and Stanton A. Glantz. "Declines in acute myocardial infarction after smoke-free laws and individual risk attributable to secondhand smoke." *Circulation* 120.14 (2009): 1373–1379.

Linnaeus, Carolus. *The System of Nature*. 1735. Digitized on Google Books, 17 Dec. 2008, from Carl von Linné, *Systema Naturae* (1756), original at Bavarian State Library.

———. *The Foundations of Botany*. 1736. Digitized on Google Books, 22 Sept. 2009, from Carl von Linné, *Fundamenta Botanica* (1741), original at Bavarian State Library.

Littlewood, Anthony, Henry Maguire, and Joachim Wolschke-Bulmahn. *Byzantine Garden Culture*. Washington, DC: Dumbarton Oaks Research Library and Collection, 2002.

Loh, Jonathan, ed., World Wildlife Fund, United Nations Environment Programme, and World Conservation Monitoring Centre. *Living Planet Report, 2008*. Switzerland: WWF International, 2008.

Loh, Jonathan, et al. "The Living Planet Index: using species population time series to track trends in biodiversity." *Philosophical Transactions of the Royal Society* 360.1454 (2005): 289–295.

Lughadha, E. Nic, et al. "Measuring the fate of plant diversity: towards a foundation for future monitoring and opportunities for urgent action." *Philosophical Transactions of the Royal Society* 360.1454 (2005): 359–372.

Manniche, Lise. *An Ancient Egyptian Herbal*. London: British Museum, 2006.

McKee, Jeffrey K. "Forecasting global biodiversity threats associated with human population growth." *Biological Conservation* 115 (2003): 161–164.

McManis, Charles R. (ed.). *Biodiversity and the Law: Intellectual Property, Biotechnology and Traditional Knowledge*. London and Sterling, VA: Earthscan, 2007.

Millennium Ecosystem Assessment. *Ecosystems and Human Well-Being: Synthesis Report*. Washington, DC: Island Press, 2005.

Morimoto, Satoshi, et al. "Morphine metabolism in the opium poppy and its possible physiological function: biochemical characterization of the morphine metabolite, bismorphine." *Journal of Biological Chemistry* 276.41 (2001): 38179–38184.

Morin, Peter J. *Community Ecology*. Oxford: Blackwell Science, 1999.

Musgrave, Toby, Chris Gardner, and Will Musgrave. *The Plant Hunters: Two Hundred Years of Adventure and Discovery around the World*. London: Ward Lock, 1998.

Myers, Norman. "The biodiversity challenge: expanded hot-spots analysis." *The Environmentalist* 10.4 (1990): 243–256.

Myers, Norman, et al. "Biodiversity hotspots for conservation priorities." *Nature* 403 (2000): 853–858.

Nahin, Richard L., et al. "Costs of complementary and alternative medicine (CAM) and frequency of visits to CAM practitioners: United States, 2007." *National Health Statistics Report* no. 18 (2009).

Nantel, Patrick, Danial Gagnon, and Andrée Nault. "Population viability analysis of American ginseng and wild leek harvested in stochastic environments." *Conservation Biology* 10.2 (1996): 608–621.

Ninan, K. N. *The Economics of Biodiversity Conservation: Valuation in Tropical Forest Ecosystems*. London: Earthscan, 2007.

Nyström, Veronica, et al. "Temporal genetic change in the last remaining population of woolly mammoth." *Proceedings of the Royal Society* 277.1692 (7 Aug. 2010): 2331–2337.

Ovadia, Ofer. "Ranking hotspots of varying sizes: a lesson from the nonlinearity of the species-area relationship." *Conservation Biology* 17.5 (2003): 1440–1441.

Paterson, Allen. *The Gardens at Kew*. London: Frances Lincoln, 2008.

Patrick, Graham L. *An Introduction to Medicinal Chemistry*, 4th ed. Oxford: Oxford University Press, 2009.

Pimentel, David, and Marcia Pimentel. "Sustainability of meat-based and plant-based diets and the environment." *American Journal of Clinical Nutrition* 78.3 (2003): 660S–663S.

Pimm, Stuart L., and Peter Raven. "Biodiversity: extinction by numbers." *Nature* 403.6772 (2000): 843–845.

Pimm, Stuart L., et al. "The future of biodiversity." *Science* 269.5222 (1995): 347–350.

Raskin, I. "Role of salicylic acid in plants." *Annual Review of Plant Physiology and Plant Molecular Biology* 43 (1992): 439–463.

Raven, John A. "The early evolution of land plants: aquatic ancestors and atmospheric interactions." *Botanical Journal of Scotland* 47.2 (1995): 151–175.

Raven, John E., et al. *Plants and Plant Lore in Ancient Greece*. Oxford: Leopard's Head (2000).

Raza, Moshin. "A role for physicians in ethnopharmacology and drug discovery." *Journal of Ethnopharmacology* 104.3 (2006): 297–301.

Reade, Julian. "Alexander the Great and the Hanging Gardens of Babylon." *Iraq* 60 (2000): 195–217.

Reinhard, Karl J., et al. "Evaluating chloroplast DNA in prehistoric Texas coprolites: medicinal, dietary, or ambient ancient DNA?" *Journal of Archaeological Science* 35.6 (2008): 1748–1755.

Ren, Dong, et al. "A probable pollination mode before angiosperms: Eurasian, long-proboscid scorpionflies." *Science* 326.5954 (2009): 840–847.

Ro, Dae-Kyun, et al. "Induction of multiple pleiotropic drug resistance genes in yeast engineered to produce an increased level of anti-malarial drug precursor, artemisinic acid." *BMC Biotechnology* 8.83 (2008).

Robbins, Gwen, et al. "Ancient skeletal evidence for leprosy in India (2000 B.C.)." *PLoS One* 4.5 (2009).

Rosenthal, Joshua P. "Equitable sharing of biodiversity benefits: agreements on genetic resources." *Investing in Biological Diversity: Proceedings of the Cairns Conference, OECD 1997*.

———. "A benefit-sharing case study for the Conference of Parties to Convention on Biological Diversity." *International Cooperative Biodiversity Groups (ICBG) Program, 1998*.

———. "Politics, culture, and governance in the development of prior informed consent in indigenous communities." *Current Anthropology* 47.1 (2006): 119–142.

Rosenthal, Joshua P., and F. Katz. "Plant-based research among the International Cooperative Biodiversity Groups." *Pharmaceutical Biology* 47.8 (2009): 783–787.

Samanani, Nailish, et al. "The role of phloem sieve elements and laticifers in the biosynthesis and accumulation of alkaloids in opium poppy." *Plant Journal* 47.4 (2006): 547–563.

Santilli, Márcio, et al. "Tropical deforestation and the Kyoto Protocol." *Climatic Change* 71.3 (2005): 267–276.

Schiestl, Florian P., et al. "The chemistry of sexual deception in an orchid wasp pollination system." *Science* 302.5664 (17 Oct. 2003): 437–438.

Schoonhoven, Louis M., Joop J. A. van Loon, and Marcel Dicke. *Insect-Plant Biology*, 2nd ed. New York: Oxford University Press, 2005.

Sheridan, Cormac. "EPO neem patent revocation revives biopiracy debate." *Nature Biotechnology* 23 (2005): 511–512.

Shi, Hua, et al. "Integrating habitat status, human population pressure, and protection status into biodiversity conservation priority setting." *Conservation Biology* 19.4 (2005): 1273–1285.

Shiva, Vandana. *Biopiracy: The Plunder of Nature and Knowledge*. Boston, MA: South End Press, 1997.

Shukla, J., C. Nobre, and P. Sellers. "Amazon deforestation and climate change." *Nature* 247.4948 (1990): 1322–1325.

Singh, Seema. "From exotic spice to modern drug?" *Cell* 130.5 (2007): 765–768.

Spalding, Mark D., Corinna Ravilious, and Edmund P. Green. *World Atlas of Corals*. Berkeley, Los Angeles, and London: University of California Press, 2001.

Strebhardt, Klaus, and Axel Ullrich. "Paul Ehrlich's magic bullet concept: 100 years of progress." *Nature Reviews Cancer* 8 (June 2008): 473–480.

Swerdlow, Joel L. *Nature's Medicine: Plants that Heal*. Washington, DC: National Geographic Society, 2000.

Tallamy, Douglas W. *Bringing Nature Home: How You Can Sustain Wildlife with Native Plants*. Portland, OR: Timber Press, 2009.

Thacker, Christopher. *The History of Gardens*. Berkeley: University of California Press, 1979.

Thomas, Chris D., et al. "Extinction risk from climate change." *Nature* 427 (2004): 145–148.

Tufts Center for the Study of Drug Development. "Drug companies still under pressure to increase pace of development." 6 Jan. 2010.

———. "New approaches to R&D may prove best path for drug developers." 6 April 2010.

United Nations Environment Programme, *Global Environmental Outlook 4*. Nairobi: United Nations Environment Programme, 2007.

United Nations, Population Division. *World Population Prospects: The 2008 Revision Population Database*. New York: United Nations, 2009.

United Nations Population Fund. *UNFPA State of World Population 2007: Unleashing the Potential of Urban Growth*. New York: United Nations Population Fund, 2007.

Vane, John R., Roderick J. Flower, and Regina M. Botting. "History of aspirin and its mechanism of action." *Stroke* 21.12 (1990): IV 12–23.

van Heerden, Fanie R. "*Hoodia gordonii*: a natural appetite suppressant." *Journal of Ethnopharmacology* 119.3 (2008): 434–437.

Vieira, Roberto F. "Conservation of medicinal and aromatic plants in Brazil." *New Crops and New Uses* (1999).

Wake, David B., and Vance T. Vrendenberg. "Are we in the midst of the sixth mass extinction? A view from the world of amphibians." *PNAS* 105, suppl. 1 (2008): 11466–11473.

Wang, Mingfu, ed. *Herbs: Challenges in Chemistry and Biology*. Washington, DC: American Chemical Society, distributed by Oxford University Press, 2006.

Ward, Bobby J. *The Plant Hunter's Garden: the New Explorers and Their Discoveries*. Portland, OR: Timber Press, 2004.

Watson, Ronald R., and Victor R. Preedy, eds. *Botanical Medicine in Clinical Practice*. Wallingford, UK, and Cambridge, MA: CABI, 2008.

Wilkinsin, Alix. "Symbolism and design in ancient Egyptian gardens." *Garden History* 22.1 (1994): 1–17.

Wilson, E. O. *Biophilia*. Cambridge, MA: Harvard University Press, 1984.

Woods, May. *Visions of Arcadia: European Gardens from Renaissance to Rococo.* London: Aurum, 1996.

World Intellectual Property Organization. *Intellectual Property and Traditional Cultural Expression/Folklore.* Booklet no. 1; WIPO publication no. 913(E). Geneva: World Intellectual Property Organization, 2005.

Yeka, Adoke, et al. "Quinine monotherapy for treating uncomplicated malaria in the era of artemisinin-based combination therapy: an appropriate public health policy?" *The Lancet: Infectious Diseases* 9.7 (2009): 448–452.

Zhou, Zhiyan, and Shaolin Zheng. "Palaeobiology: the missing link in ginkgo evolution—the modern maidenhair tree has barely changed since the days of the dinosaurs." *Nature* 423.6942 (2003): 3.

Online Resources

Convention on International Trade in Endangered Species of Wild Fauna and Flora: cites.org

Fogarty International Center: fic.nih.gov

Food and Agricultural Organization of the United Nations: fao.org

International Cooperative Biodiversity Groups: icbg.org

International Union for Conservation of Nature and Natural Resources: iucn.org

IUCN Red List of Threatened Species: iucnredlist.org

National Park Service: nps.gov

The Plant List: theplantlist.org

Traditional Knowledge Digital Library: tkdl.res.in

US Department of Health and Human Services: hhs.gov

US Food and Drug Administration: fda.gov

US Fish and Wildlife Service: fws.gov

World Health Organization: who.int

World Intellectual Property Organization: wipo.int

Index

About the Author

Kara Rogers is the senior editor of biomedical sciences at Encyclopaedia Britannica, Inc. She holds a BS in biology and a PhD in pharmacology and toxicology and is a member of the National Association of Science Writers. When not writing or reading about science, she loses herself sailing the waters of Lake Michigan, tending the grapevines on her parents' Kentucky vineyard, or watching birds and hiking and camping. She and her husband, Jeremy, live in Chicago.